PRAISE FOR
Surviving Climate & Chaos

"When there is a global effort by the reckless right to promote fossil fuels over renewable energy, Evan Jevnikar's *Surviving Climate and Chaos* is a cautionary tale of how another time the atmospheric CO_2 levels rose to exceptional levels, the result was disastrous—including for the mighty dinosaurs *Tyrannosaurus* and *Triceratops* among them."

—**Dr. Gregory Paul,** author and illustrator of
The Princeton Field Guides to Dinosaurs

"In accessible language, Evan Jevnikar narrates how life on our planet evolved and changed in response to changing climates over the last half billion years. By deftly weaving between ancient environments and their plant and animal inhabitants and the present day, Jevnikar demystifies the science behind climate change by explaining its underlying drivers and ancient roots using relatable metaphors and examples. This is an engaging and fun read for anyone curious about Earth's geological, biological, and climate history, and how those bear on present-day challenges."

—**Dr. Peter Makovicky,** professor at University of Minnesota

"A brilliant blend of adventure and evidence. Evan Jevnikar makes the age of dinosaurs speak directly to today's climate crisis. A must-read for all lovers of prehistoric life!"

—Joshua Jung, paleoartist (@Ingen_fx)

"A fascinating blend of deep-time paleontology and modern climate science. Jevnikar shows us the past to help us navigate the future. There is hope for a better tomorrow if we act today."

—Dr. Brian Curtice, founder of Fossil Crates

"A lively account of how changing climates have influenced the history of life on Earth."

—Dr. Mike Benton, professor at University of Bristol and author of *Extinctions: How Life Survives, Adapts and Evolves*

Surviving Climate & Chaos

Surviving Climate & Chaos

What Dinosaurs Teach Us About Climate Change and Resilience

By

EVAN JEVNIKAR

🥭 mango
PUBLISHING

MIAMI

Cover, Layout & Design: Megan Werner
Cover Illustration: Алексей Воробьёв, ZinetroN / stock.adobe.com
Interior maps: https://climatearchive.org/about.html
Interior dinosaur charts: Andy Wilson, Christine Axon, dannj, François-Louis PELISSIER, Itai Fein, Jagged Fang Designs, JFstudios, Manuel Brea Lueiro, Matías Muñoz, Mike Taylor, Pranav Iyer, risiattoart, Robert Gay, Scott Hartman, Steven Traver, Tasman Dixon, thefunkmonk, T. Michael Keesey / phylopic.org

For permission requests, please contact the publisher at:
Mango Publishing Group
5966 South Dixie Highway, Suite 300
Miami, FL 33143
info@mango.bz

For special orders, quantity sales, course adoptions and corporate sales, please email the publisher at sales@mango.bz. For trade and wholesale sales, please contact Ingram Publisher Services at customer.service@ingramcontent.com or +1.800.509.4887.

Surviving Climate & Chaos: What Dinosaurs Teach Us About Climate Change and Resilience

Library of Congress Cataloging-in-Publication number: 2025942900
ISBN: (pb) 978-1-68481-869-3 (e) 978-1-68481-870-9
BISAC category code: SCI054000 SCIENCE / Paleontology

Printed in the United States of America

For my wife, McKenzie, and our children

Table of Contents

INTRODUCTION

Answers to the Future Found in the Past

I set my pack down and pulled out my tools. I grabbed my small screwdriver and my rock hammer and gently began chiseling away the rock. It wasn't long before I found something hidden in the rock: a small thigh bone about the size of my hand. As I tried to dig around the bone, I ran into small toe bones, ribs, and back bones. These didn't belong to any of the desert animals that lived here in Utah. The unique shape of these bones revealed that they belonged to something more ancient: a dinosaur. More specifically, these bones belonged to a species of dinosaur called *Falcarius*. It was a roughly human-sized dinosaur that walked on two legs, had a long neck, and sported hooked claws on its hands. Far from the ferocious *Velociraptor* seen in *Jurassic World*, *Falcarius* was a gentle herbivore that probably used its hooked claws like a sloth to pull down tree branches and eat leaves and fruit.

After digging under the hot sun for about two weeks, I managed to dig one meter (three feet) into the hillside. Just in that small section, there were nearly a hundred *Falcarius* bones all stacked on top of each other. It was the same for the rest of the team I was digging with; by the end of the trip, we had collected nearly a thousand separate bones. Altogether, it summed up to several individuals of *Falcarius*, all ranging in size from adults to tiny little babies. While this was my first time digging in this quarry, my mentor had been digging here for nearly two decades. Each year produced similar results. It was a mass grave site for these prehistoric animals. Why did all of these *Falcarius* die in this spot?

We can begin to understand by gathering context clues from the fossil quarry. The rock that all these bones were found in was very fine-grained; if you looked at it under a microscope, it would look like mud that had been compacted over time. When I was digging up the leg bone, it wasn't oriented horizontally, as if it was lying down; it was actually straight up, as if it was sinking. Both of those clues seem to suggest that this used to be a swampy area. Therefore, the reason all of these *Falcarius* were found here is because individuals likely wandered into this swampy area over the years, and many of these dinosaurs got stuck, died, and were buried within the swamp.

Utah isn't known for its swamps—it's primarily a desert. It's a hot and dry desert, like many of the other Southwestern states in the US. Utah was a completely different landscape when *Falcarius* was alive. Swamps and deserts require completely different climates to form. So, the climate *Falcarius* lived in is completely different than the one we are currently living in. This isn't just the case in Utah. In New Mexico, next to the dinosaur fossils I dug up were the fronds of palm trees and little nuggets of amber. In landlocked Kansas, along the limestone chalk that dots the highways, I've found coral, sea sponges, and teeth of fish that would have been the size of a sedan. In the badlands of Montana, where we find the monstrous T. rex and the imposing Triceratops, we find seeds of giant pine trees, strips of bark, and petrified stumps of trees you'd normally find in temperate forests. Wherever you find dinosaur fossils, you are also likely to find evidence of a climate totally different than the current one. Even by comparing quarries that represent different time periods, with different dinosaurs, we can see that their climates too were different and changed over time. All of this is to say that the climate of our planet is not a fixed feature. It changed across the reign of the dinosaurs and has changed since their extinction.

Climate change is one of the big buzzwords of today. You probably hear a lot of things about it. You hear it in the news, from our politicians, and from companies. Depending on who you talk to, the responses on climate change range from "It's destroying our planet" to "It's not real." So, what's the deal? How can we make heads or tails of such a divisive topic?

Just from fossil evidence, it looks like it's not fake. Not only that, it has a long history on our planet from way before human societies popped up. Dinosaurs have lived through plenty of climate change during their reign over prehistoric Earth. In fact, they are probably one of the best case studies for understanding climate change. Why? Well, there are a couple reasons. One is that dinosaurs, as a group of animals, have existed for a long time.

Looking at how modern animals adapt to changing temperatures in history might help, but not enough time has passed. There are 180 million years between the first dinosaurs ever to exist and their extinction in the Cretaceous Period, sixty-six million years ago. *Dinosaurs existed for longer than they have been extinct.* Another reason is that dinosaurs lived on every single continent: You can find dinosaur fossils from the North Pole to equatorial South America, and down to the South Pole. From the western hemisphere to the eastern hemisphere, they literally lived everywhere. As such, they carved out a living in every type of environment you could think of. Overall, they were a very hardy group of animals. While some dinosaurs would go extinct as their environments changed, others survived and adapted to that change.

One dinosaur isn't going to tell us much about their evolution or climate change. However, two different species of dinosaurs might. What about more than two? Over the past two hundred years of research in the field of paleontology, we've discovered thousands of species that lived within this 180-million-year period. Just in 2024 alone, we've found over fifty new species, and we're finding more every year. With all of these species—all of these unique data points—we can get a pretty comprehensive picture of what dinosaurs were like back then.

We can track some pretty interesting changes over time using two hundred years of dinosaur discoveries. We can look at basic things, like how they changed size or how the shape of their bones changed over time. Yet, now we can understand even more intimate things, like how their growth and metabolism changed, how their diets changed, and what kind of soft tissue structures they developed. When we track these changes in dinosaur physiology with changes in temperatures, we can get a unique perspective on how climate change affected dinosaurs.

"But why should we care about animals that lived millions of years ago? The fact that T. rex existed sixty-six million years ago doesn't affect my life."

The existence of T. rex probably won't change your life. However, the entire sum of paleontology research so far can teach us something. That's the whole point of science: using past experiences and observations to be able to predict and better understand the future. With this wealth of data that's been collected over time, we can get a really deep understanding of how plants, animals, environments, and the weather are affected by climate change. From there, we can accurately predict what the future of climate change looks like here on Earth.

So, to understand a little bit more about climate change, we're going to look back in time to the dinosaurs to see how they lived through climate change. We'll take a trip to their humble beginnings in the Early Triassic, their meteoric rise over time, and their catastrophic extinction. Over 180 million years, we can begin to ask questions like, "What caused climate change? How different were global or even regional temperatures? How did environments change due to climate change? How did dinosaurs adapt to this climate change?" And most importantly, "Who went extinct and why?" After we understand climate change of the past, it will become clear what lies in store for us in our current state of climate change. As we'll see, dinosaurs lived through some pretty harrowing periods of intense climate change. However, the solutions to our own climate crisis may actually surprise you.

PART 1

End Permian and Early Triassic: *The End of the World*

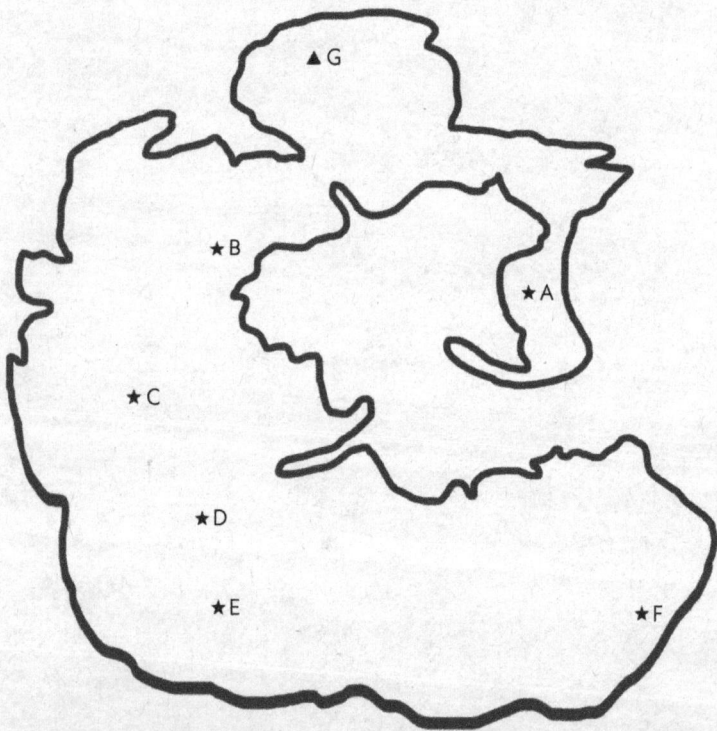

Pangea: ~251.9 million years ago

A - Beijing   C - New York City   E - Johannesburg   G - Siberian Traps

B - London   D - Rio de Janeiro   F - Sydney

Origin of life
Invasion onto land
A First forests
End-Permian extinction

Geologic Timescale

Rise of the mammals
The Ice Age B
Modern history

A ⟵ | Mesozoic | ⟶ B

| Triassic | Jurassic | Cretaceous |

1 2 3 4 5 6 7 8

251 246 237 201 174 161 143 100 66

Time | Milllions of Years Ago

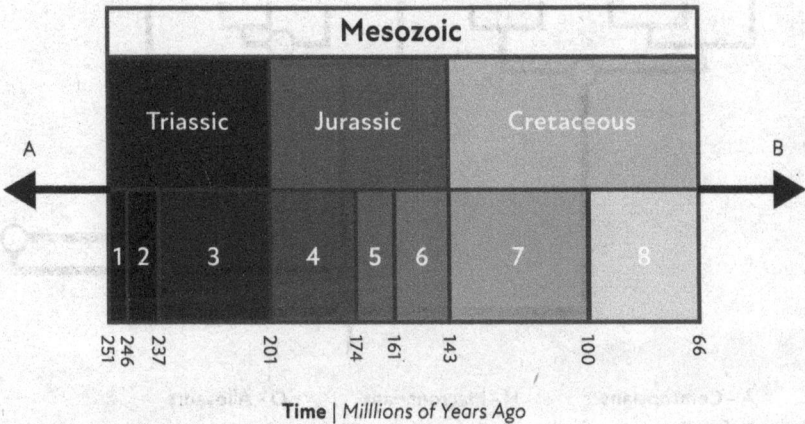

1 - Early Triassic   3 - Late Triassic   5 - Middle Jurassic   7 - Early Cretaceous
2 - Middle Triassic   4 - Early Jurassic   6 - Late Jurassic   8 - Late Cretaceous

Prehistoric Animal Family Tree

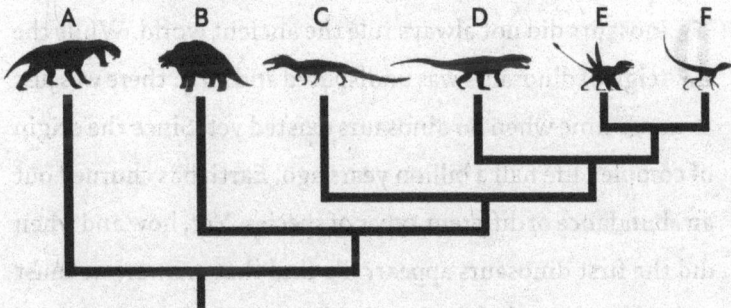

A   B   C   D   E   F

A - Therapsids   B - Reptiles   D - Rauisuchians   E - Pterosaurs
    Dicynodonts   C - Early Archosaurs   Aetosaurs   F - Dinosaurs
    Mammals   Crocodiles

Dinosaur Family Tree

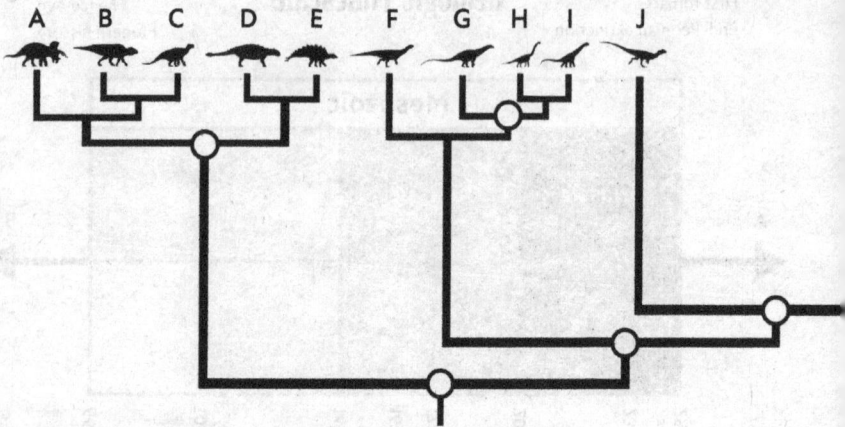

A - Ceratopsians H - Macronarians O - Allosaurs
B - Hadrosaurs   I - Titanosaurs   P - Carcharodontosaurs
C - Iguanodonts   J - Early Theropods Q - Tyrannosaurs
D - Ankylosaurs   K - Ceratosaurs   R - Oviraptorosaurs
E - Stegosaurs   L - Abelisaurs   S - Dromaeosaurs
F - Sauropodomorphs M - Megalosaurs   T - Birds
G - Diplodocids   N - Spinosaurs

Dinosaurs did not always rule the ancient world. While the reign of dinosaurs was undisputed and long, there was just as much time when no dinosaurs existed yet. Since the origin of complex life half a billion years ago, Earth has churned out an abundance of different types of species. Yet, how and when did the first dinosaurs appear? To find that answer, we must go back to the end of the Permian Period. The Permian Period was as foreign to the dinosaurs as the Cretaceous Period is to us. Even though it was 252 million years ago, many features of

K   L   M   N   O   P   Q   R   S   T

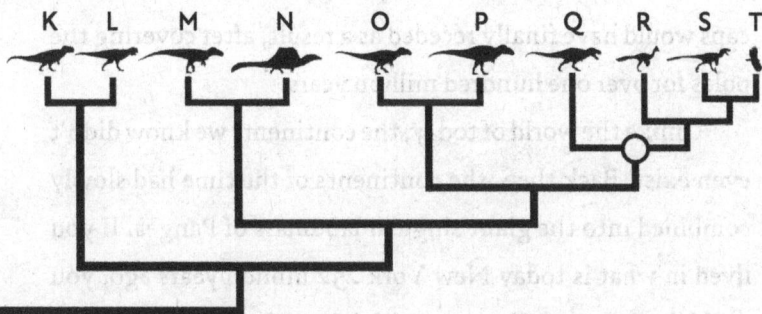

End Permian & Early Triassic Animals

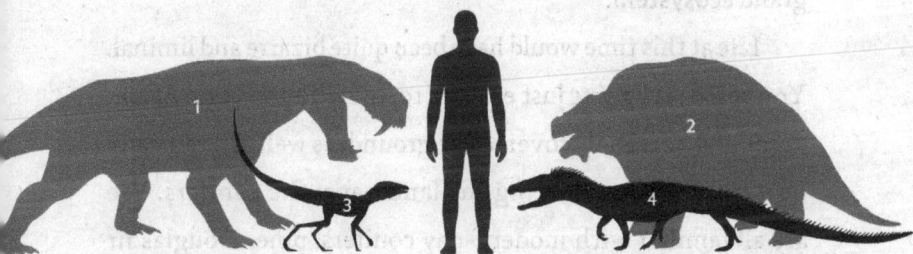

1 - *Inostrancevia*, a gorgonopsid
2 - *Scutosaurus*, a pareiasaur
3 - *Asilisaurus*, one of the earliest ancestors of dinosaurs
4 - *Archosaurus*, an archosaur

the end of the Permian mirror what we see in today's modern age. As we'll see, what led to the origin of the dinosaurs was far from a comfortable and easy beginning.

Similar to recent history, Earth was in the middle of an ice age. However, this earlier ice age was brutally cold, colder than the ice age woolly mammoths and saber-toothed cats lived through. This previous ice age saw the first woody trees evolve, but now dense forests and swamps began to fade away. The ice caps would have finally receded as a result, after covering the poles for over one hundred million years.

Unlike the world of today, the continents we know didn't even exist. Back then, the continents of the time had slowly combined into the giant singular landmass of Pangea. If you lived in what is today New York 252 million years ago, you could, in theory, get in a car and drive to what is now Morocco in a couple of hours. Given that all land was connected, all life was free to move and migrate across Pangea and create a grand ecosystem.

Life at this time would have been quite bizarre and liminal. You could recognize just enough to know you were on Earth. Ferns and horsetails covered the ground, as well as a radically new type of plant dotting the landscape: the conifers. We are all familiar with modern-day conifers: pine, Douglas fir (Christmas trees), junipers, and the giant redwoods. However, these prehistoric plants were much smaller than modern-day conifers. Neither were they prominent enough to create vast forests. Many early conifer species were just getting familiar with their environments. Bugs reminiscent of dragonflies and damselflies would have been buzzing around. As for the other animals that made up most of the Pangean ecosystems, they would have been entirely foreign to what we would recognize today.

These animals are known as the therapsids. If you saw a therapsid today, you might be confused as to what type of animal it was. Therapsids looked quite alien: nothing like anything that lives today. These animals had legs that were semi-erect, meaning that they would have stood slightly bow-legged, but not to the extent that lizards sprawl out. Most of them would have been covered in leathery skin like an elephant, but some may have had small patches of hair, or even whiskers. A vast majority of them also sported big toothy or tusked skulls. Despite this, these animals were very ancient ancestors to mammals, though you could hardly tell. During the Permian, therapsids dominated the landscape. However, today, they are nowhere to be found.

If you were to zoom into the Permian environment a bit more, you would likely see it populated with therapsids called dicynodonts. These herbivores ranged in size from a small dog to (roughly) a grizzly bear. Their large bulky heads held a sharp beak, used to shear plants, and massive jaw muscles that helped them chew these plants. Due to their lowly stature, they would only have been able to manage to eat food close to the ground. A pair of tusks protruded from the beak as well. As with modern tusked animals, these may have been used to fight off rival individuals to secure mates, or defend what little land they had available to them in the shadows of much larger predators.

At the apex of the therapsids was *Inostrancevia*, the dominant land predator of Siberia at the time. This ancient animal was about the size of a tiger, but is believed to have had no hair. Its head was long and robust, featuring a pair of large saber teeth, similar to the saber-toothed cats of the most recent ice age. These large bladed teeth would have been efficient at killing prey. Based on similarly-sized relatives, it's estimated that *Inostrancevia* had a bite force of 715 newtons, roughly equivalent to the bite force of a German shepherd today. While relatively weak by modern standards, *Inostrancevia* was the pinnacle of predator evolution at this point in the Permian.

Shifting the focus to other animals of the Permian, amphibians and reptiles played their parts in the Pangean environments, albeit to a lesser extent than therapsids. One of the more notable reptiles was *Scutosaurus*, a rotund pony-sized reptile. Its massive size made it slower than most therapsids, capable of only short bursts of speed. Even so, it more than made up for this with its state-of-the-art body armor. *Scutosaurus* was covered head to toe in tough bony armor similar to the scutes of crocodiles today. In addition, its cheeks and chin sported multiple large spikes that it would have slammed into any predator that made a move for its neck. Any predator that managed to catch *Scutosaurus* would have to fight for its meal.

Amongst this diverse group of strange animals was a particularly unique character, and it is where our story of the dinosaurs begins. Up in Siberia, this three-meter-long (or ten-foot-long) scaly animal crawled on all fours in search of unsuspecting prey. Its name was *Archosaurus*. The face of

Archosaurus was imposing: Like a crocodile, it bore its many large teeth for all to see, without any lips hiding them. At the front of its snout, in its upper jaw, was a notch, allowing for its snout tip to curve downwards, resulting in a sort of hook in its upper jaw. This caricature-like feature was actually *Archosaurus*'s deadliest weapon. When it caught smaller animals in its mouth, this hooked upper jaw was efficient at keeping them stuck in their jaws, making it impossible for their prey to wriggle free. While menacing by human standards, *Archosaurus* was not an apex predator of its time, so it was forced to hunt smaller prey. Yet, while *Archosaurus* was not at the top of the food chain in the Permian, this would change for its descendants.

If you saw *Archosaurus* today, you'd probably mistake it for an exotic type of crocodile, but it was actually more ancient than both crocodiles and dinosaurs. Crocodiles (as well as alligators, caimans, and gharials) and dinosaurs belong to the same group of animals, called archosaurs, of which *Archosaurus* is a close ancestor. This means that crocodiles are not a type of dinosaur, but are some of the closest living relatives to dinosaurs. Thus, the story of dinosaurs begins with these ancient ancestors. *Archosaurus* was not merely special for its unique skull shape, or because it looked similar to a dinosaur. Underneath its scaly exterior was its true latent superpower, found in all archosaurs.

You and I, and all mammals for that matter, have bidirectional lungs. We breathe in and then we breathe out.

We absorb oxygen from the air into our lungs, and then release carbon dioxide as a byproduct afterwards. Once we release all the carbon dioxide, our lungs are empty, requiring us to start the cycle of breathing all over again. This is how all of the dominant therapsids of the Permian breathed. Not archosaurs, though.

Based on their modern relatives, all ancient archosaurs likely had unidirectional lungs, which are completely different. As it breathed in, air would first flow through air sacs that would act as a storage center for this fresh unused air. This air would then funnel into lungs, where it would then be absorbed, and carbon dioxide would be released. Before it was fully released, it would move to a second air sac, allowing any oxygen that hadn't been absorbed yet to be absorbed.

If you hold your breath, you probably can only do it for a minute or so before you have to breathe out. Carbon dioxide can't be absorbed, meaning we need to get it out of our lungs to make space for oxygen. It's more important for us to release the carbon dioxide in our lungs than extract every bit of oxygen. Therefore, we breathe out some oxygen that hasn't been used yet. Our current atmosphere has plenty of oxygen, so we don't have to worry about our technically wasteful breathing; there's plenty of oxygen to go around, and not so much carbon dioxide.

Since its air sacs would act as a temporary storage center for air, this meant that *Archosaurus* always had fresh air ready for its body to absorb. The second air sac, after its lungs, gave it a chance to fully absorb the oxygen in its lungs before breathing out. Its breathing was hyper-efficient, giving it ultimate stamina in its environment. It would be impossible for our lungs to even come close to what *Archosaurus* could do naturally. Unidirectional breathing gave *Archosaurus* and other early archosaurs a leg up over any other animal in its environment. It could not compete with the raw power of the gorgonopsids, but this meant it could chase down prey with ease.

Overall, the end of the Permian boasted many unique animals that set the foundation for life today. As these ecosystems transitioned out of an ice age, it seemed that life was heading toward a climate very similar to what we have today.

But this could not be further from the truth.

252 million years ago in Siberia, deep within the Earth, between the core and the mantle, a massive plume of magma began to rise to the surface, known as the Siberian Traps. This plume was so big that instead of a single volcano emerging, the whole landscape became filled with volcanoes and created a large igneous province. The sheer size and scope of this volcanic province is unlike anything we know of today on Earth. Eventually, the magma underneath all of these volcanoes reached a breaking point, and all hell broke loose. Lava similar to the runny and flowy lava found in Hawaiian volcanoes rapidly flowed across the land. For a

million years straight, a deluge of fifteen million kilometers3 (3.6 million miles3) of lava flooded over Siberia. It covered seven million kilometers2 of land and was anywhere from one to 6.5 kilometers (0.6 to four miles) deep. All the land from Kostanay, Kazakhstan, to Yakutsk, Russia, had been drowned in lava. This was the largest volcanic eruption that has ever happened in Earth's history. Moreover, it sent temperatures around the globe skyrocketing.

Just one volcano can't warm our entire planet. For example, the eruption of Mt. Pinatubo in 1991, which was the second-largest eruption of the twentieth century, didn't warm the planet (in fact, the sulfur dioxide it released actually cooled the planet). Even a single supervolcano, like Yellowstone, wouldn't be able to warm the planet if it erupted. Only a massive province of volcanoes, like those in Siberia, all erupting at the same time could produce enough lava and release enough carbon dioxide to actually change the climate. With the Earth coming out of an ice age, it was all too vulnerable to things that would cause global warming. This precarious state is one that we seem to find ourselves in during our modern age as well.

Under normal circumstances, this event would have eradicated a whole hemisphere of life and set Earth on a course toward global warming. This alone would have been bad enough, but this was merely the prelude. During the preceding ice age, plants had begun to develop one of the most essential features we see today: wood. This plant feature was a novel biological trick, in that it kept trees protected, given that they were stuck to the ground. However, it was so novel that no microorganisms had yet evolved to break down wood. Thus, these first trees did not decompose. Instead of breaking down into soil, whole swaths of land filled with dead but intact trees were buried. Eventually, bacteria and other organisms evolved to break down wood. Even so, all of those trees that never decomposed were buried and compacted over millions of years. The carbon that had been absorbed by these trees got condensed to form coal—the same type of coal we use today. These global forests had produced millions of square miles of coal just underneath the surface during the Permian. When this sea of lava melted into these coal deposits, it scorched millions of tons of coal, producing an enormous amount of carbon dioxide, similar in effect to humans burning fossil fuels today.

Why does carbon dioxide lead to global warming? Carbon dioxide is a greenhouse gas, unlike oxygen, the substance that humans and other animals breathe in from the air. The unique structure of carbon dioxide allows it to absorb a lot of energy from infrared radiation. As a result, carbon dioxide can release that extra stored energy, in the form of heat, back into the atmosphere. In an atmosphere with less carbon dioxide, infrared rays mostly pass through the atmosphere, warm the surface, and then head back out through the atmosphere. However, an atmosphere with more carbon dioxide absorbs the heat from infrared radiation. Carbon dioxide then reemits the heat in all directions; this heat is sent back to the surface of the Earth, where it warms the surface of the planet again. This constant absorption and reemission process caused by carbon dioxide allows much more heating of the Earth's surface to occur. In fact, the biggest predictor for the temperature at any point in Earth's prehistory is the proportion of carbon dioxide in the atmosphere, which can be calculated from fossil proxies.

There will always be a certain amount of carbon dioxide in the atmosphere, and that's okay. Animals naturally produce carbon dioxide when they breathe, meaning that it will always be here, and it always has been here. It's completely normal, and even beneficial to our atmosphere. In fact, having barely any carbon dioxide and an abundance of oxygen in the atmosphere can lead to an ice age, which is exactly what happened before the Permian. The atmosphere has to balance each of these two substances, allowing for minor variation from time to time, to keep the climate stable. Now it becomes clear why the mass burning of coal could only exacerbate the already intense climate change of the End Permian. The burning of all this coal in the span of about a million years, in addition to the massive eruptions of lava, released roughly twenty-six million gigatons of carbon dioxide into the atmosphere in under a million years. That may seem like an enormous amount when viewed in its entirety—and it is—but that is only an estimated twenty-six gigatons of carbon dioxide on average every year.

For perspective, roughly 36.8 gigatons of carbon dioxide was released into the atmosphere just in 2023. Yet in the span of a year, we hardly noticed a difference in our world from the addition of 36.8 gigatons. It's only over the course of nearly a century of burning fossil fuels that we are starting to see the effects; from the mid-nineteenth century to now, temperatures have increased, deserts have expanded, and weather events have become more severe. However, in our current climate crisis, we have the ability to change it and take steps toward solving this rapidly growing issue. There are many ways societies across the globe can redirect global warming. In nature, this is accomplished passively by our natural carbon sinks—parts of the environment that naturally absorb carbon dioxide.

Nature has been surviving on this planet much longer than we have, and has figured out a way to create an atmospheric balance that works for nearly all life. Plants take in both light and carbon dioxide and convert them into breathable oxygen for animals, who in turn breathe out carbon dioxide for plants. As plants die and break down, the carbon dioxide in the plant material is taken in by the microorganisms and bacteria and stored in the soil. Breaking down the other organic material of plants creates readily available nitrogen and carbon for plants to use to grow. It creates a positive feedback loop: Healthier soil makes it easier for trees to grow. The more trees that grow, the more carbon dioxide is absorbed. The more trees that break down into soil, the more carbon dioxide is absorbed, and the healthier the soil becomes. This makes forests one of the most important engines for actually *reversing* global warming. Even though the change we currently face in our climate is similar to what was happening during the Permian, humans can prevent it. However, in the Permian, no one could stop the immense burning of coal or the vast deforestation caused by the volcanic eruptions.

The end of the Permian saw these side effects of global warming taken to their extreme conclusion. Instead of an ice age, a hothouse climate emerged. Hothouse climates usually have an average temperature above 20 °C (68 °F) with the poles reaching up to 5 °C to 15 °C (41 °F to 59 °F), preventing ice from accumulating. However, this hothouse climate was far more deadly. There was nowhere safe on Pangea as temperatures skyrocketed from 17 °C (62.6 °F) to an average of 32 °C (89.6 °F). The poles, which were normally below freezing most of the time, rose to an average of 15 °C (59 °F). This made it impossible for ice to accumulate, resulting in the rapid melting of the ice caps. Pangea turned into an unlivable nightmare. The effects would have hit the hardest in the center of Pangea. If you were dropped in the middle of Pangea at the peak of the hothouse, you would have no chance of surviving, or even escaping toward anything remotely hospitable.

Since Pangea was a giant landmass, it was much more susceptible to drying up in the center and creating deserts there. Under these circumstances, the interior of Pangea broiled at an average annual temperature of 50 °C (122 °F). The constant heat dried up all the soil for miles. This made it impossible for water to be retained or plants to grow, causing deforestation in a different way, and even further impacted the natural intake of carbon dioxide into the environment. As soil dries out, the microorganisms that break down the carbon dioxide die. Thus, no nutrients are then available for plants to grow. With no topsoil in these deserts, no microorganisms or plants could survive. Without these essential components taking carbon dioxide out of the atmosphere, Earth spiraled out of control.

This barren desert landscape stretched out for thousands of miles in all directions. An area from modern-day western Bolivia to the east coast of Brazil, plus the coast of Nigeria to the southern border of Libya, had all withered away to a scorched landscape. To call it a wasteland would be inaccurate. It was hell on earth.

Life stood no chance. With temperatures rising so quickly, there was no time for animals or plants to adapt to such hot conditions. Thus, the Earth went through one of its worst extinctions in history. 90% of all species died off, and it's estimated that up to 99% of all individual organisms died. To put that into perspective, imagine if Manhattan—a New York City borough of 1.6 million people—had its population struck down to sixteen thousand people. Any survivor would likely think they'd lived through the Rapture.

All the other therapsids that lived nearby, like *Inostrancevia*, stood no chance of survival at the epicenter of this destruction. In fact, all therapsids were wiped out, except for a few species of dicynodonts and small gorgonopsids. These meager survivors lived on, but at a fraction of their peak diversity. *Archosaurus* went extinct as well, since it lived right in the middle of the lava flows. However, *Archosaurus* was just one of many early archosaurs that lived throughout Pangea, many of which were farther away from these lava flows. As nearly all other prehistoric animals died off, the archosaurs as a group survived. Why did the archosaurs survive when nearly all other animals died off?

Remember, the archosaurs had hyper-efficient lungs that were excellent at sucking up oxygen from the air. Now, with carbon dioxide dominating the composition of air, there was relatively little oxygen to breathe in. For the "wasteful" breathers, like the therapsids, this was a death sentence, as they struggled to simply breathe. If you've ever hiked up a mountain at high altitude, you'll know how much harder it is to exercise with less oxygen. Now, imagine if your life depended on chasing down food or fighting off predators in this environment. It would be extremely difficult. Archosaurs, though still struggling, were able to make do thanks to their unidirectional lungs.

This was not all. In an environment where water was not guaranteed, every drop counted. Even excreting water could cost you your life. Therapsids, like us, peed out urea. This byproduct is quite soluble in water and requires water for excretion. However, archosaurs peed out uric acid, which is less soluble and can be excreted in a nearly solid form, with very little water. Archosaurs were able to conserve more water in their bodies and thus dehydrate at a much slower rate, giving them another advantage over therapsids.

Finally, because therapsids were the dominant land animals, they had an abundance of opportunities in their ecosystems. When such opportunities were presented to animals in the past, many species grew bigger and/or became more specialized in their diets. History is replete with cases like this. If they were large, they required more nutrients, and thus a greater volume of food. If they were more specialized, they became more picky eaters. In an apocalypse situation like this, these characteristics never fare well. Since the Earth was heating at such a rapid pace, plants dried up relatively quickly. Thus, there was less plant material for herbivores to eat, resulting in herbivores starving. The bigger a species was, the more plants it needed, and the more likely it was to die off and go extinct. If a species was unlucky enough to only dine on a select few plants before the extinction, this event meant they were out of options and promptly went extinct. With many species of herbivores dying off, there was little left for the carnivores to eat. Faced with the same situation, many carnivores starved as well, with the biggest and most specialized going extinct first. As more and more species went extinct, the food webs, composed mostly of therapsids, across Pangea quickly disintegrated.

All of these features gave the archosaurs an advantage in this new heat- and water-stressed world. After the Siberian Traps, archosaurs began to populate the world. Despite their advantages, Pangea at the beginning of the Triassic was still unlivable in many places. Many archosaurs spread to havens like Europe, southern Africa, and Argentina, which were situated far from the hostile interior. It was here that archosaurs had a biological boom. Before the Siberian Traps erupted, archosaurs were relatively rare and small. After the eruptions subsided, there were many new species, of all shapes and sizes. Some, like the massive *Erythrosuchus*, rivaled Nile crocodiles in size.

Within the hustle and bustle of the Early Triassic archosaur boom was a humble type of archosaur that occupied the lowest rung of the food chain. It was small and nimble. It left footprints unlike anything seen before from an archosaur. Like the crocodiles and alligators of today, many archosaurs walked flat-footed; when the feet of crocodiles and alligators land on the ground, all their toes and their heels make contact with the ground. This way of walking is called plantigrade walking. In fact, humans walk this way as well; our heels touch the ground. Yet these mysterious footprints were different because this archosaur put more pressure on its toes when it walked, leaving the heel slightly elevated off the ground, the outer toes barely touching the ground. This would be equivalent to our big toe and pinky toe. Imagine walking on just your inner three toes.

These unique footprints belong to some of the earliest relatives of dinosaurs—essentially pre-dinosaurs—called *Prorotodactylus*. This animal lived in what is now Poland and France 249 million years ago, only a few million years after the cataclysm of the Siberian Traps. This seems to suggest that the precursors to dinosaurs were small, agile animals that would have been a far cry from their magnificent descendants. How is it that such a small animal was able to survive one of the worst mass extinction events in history? The precursors to dinosaurs being relegated to lower rungs on the food chain was actually a saving grace. On average, they were much smaller, and had a generalist diet in order to survive in such a competitive position in their environment. This meant they could survive with less food, and could eat most of the food that came their way. While many larger species of archosaurs still died off, the smallest and most adaptable relatives pushed through the starvation and lived long enough to pass on their genes. Additionally, the smaller an animal is, the easier it is for it to find places to hide during stressful periods. These animals could hide in burrows, underneath logs, or behind rocks to keep themselves safe, while larger animals were forced to weather whatever storm came their way. The phenomenon of small animals overwhelmingly succeeding in the wake of mass extinctions is called the Lilliput effect. This meant that the tiny precursors to dinosaurs were the ultimate doomsday preppers: able to hold out while conditions were bad and survive on little to no food.

Not long after we see these footprints, around 245 million years ago, we begin to see what many consider the first dinosaur: *Asilisaurus*, a type of early dinosaur called a silesaur. This little dinosaur was no bigger than a golden retriever; it was roughly one meter (three feet) long and 0.6 meters (two feet) tall. It would only have weighed about one kilogram (twenty-two pounds). *Asilisaurus* had a long neck and erect limbs, meaning that their legs didn't sprawl out like a lizard's.

What makes these early dinosaurs so significant? While their small size helped them survive a mass extinction, they were likely snacks for even bigger and more powerful archosaurs. Dinosaurs had all the standard features of an archosaur that helped them navigate the scalding hellscape of Pangea: unidirectional lungs for effective breathing, efficient water conservation, and erect limbs for better locomotion. However, dinosaurs had an even greater suite of secret weapons.

One was their erect posture. Even though archosaurs were already efficient breathers, there's one major issue with having a sprawling posture: An animal can't breathe and run at the same time. Each time a lizard or crocodile takes a step, they are actually constricting one side of their lungs. As a result, most reptiles are only capable of short bursts of energy, simply because if they run for too long, they'll pass out. This is not the case for animals with an erect posture, like early dinosaurs. Their lungs didn't constrict as they took a step, which meant they were one of the few animals at the time that could breathe and run simultaneously. Combine that with the hyper-efficient lungs of the archosaurs, and you had an animal with unmatched stamina. Early dinosaurs, like *Asilisaurus*, could run to get away for a longer time if they were being chased by a predator. They could also chase after smaller prey for longer, and likely outpace them.

Another advantage was that their hips were actually quite similar to humans'. Our hips have a socket that allows our thigh bone to fit in directly. This helps provide stability when running, keeping the leg oriented forward and making it very hard for it to wiggle sideways. Dinosaurs were unique among archosaurs in that they too had hip sockets. Yet, unlike us, another special feature was their powerful tail muscles. All dinosaurs had a powerful muscle that attached from their tail to the middle of their thigh bone. This muscle acted as a powerful contracting force when running. As a dinosaur took its step forward, the tail muscle then contracted and pulled the leg back, helping dinosaurs decrease the time it took for them to take a step, thus increasing their speed.

Finally, dinosaurs moved away from walking on their heels and began walking on their toes. At a walking pace, this wouldn't make much of a difference, but all elite sprinters know that running on your toes is the perfect form for speed. When you sprint on your toes, there is less surface area of your foot making contact with the ground. This reduces the impact force of running. Additionally, pushing off just your toes when you start sprinting, instead of your whole foot, takes less time and already engages the leg muscles needed to sprint. This would come in handy for an animal that was being ambushed and had little time to accelerate away from a predator. Not only were early dinosaurs scrappy survivalists; they were impeccable athletes, too.

Put all of these advantages together, and dinosaurs were the quickest, longest-lasting, and most agile runners ever to exist on Earth up to this point. No animal of the same size stood a chance. In a hostile environment where food and water were scarce, every step counted. You needed to be a marathon walker. You also needed to avoid being a meal with deadly predators waiting around every corner. These dinosaurs could do it all. We often think that because evolution means survival of the fittest, the biggest and strongest survive. However, "fittest" doesn't always mean athletically fit. A better way to describe it would be like the way a puzzle piece fits together with the other pieces around it. In this sense, dinosaurs were some of the "fittest" puzzle pieces within the environments of the Early Triassic.

Eventually, the destruction brought by the End Permian subsided. In the Middle Triassic, average global temperatures cooled to roughly 24 °C (75.2 °F). This was still much hotter than today, but it was much colder than the peak of the end-Permian extinction and Early Triassic. The hothouse climate was over. This little bit of cooling allowed a few surviving plants to rebound that were already adapted to handle hot and arid conditions before the disaster struck. These plants were the conifers and a unique type of plant that no longer exists today, known as seed ferns; seed ferns look very similar to modern ferns with their frond-like branches. However, ferns require water in their immediate surroundings for the spores they release to find each other and reproduce. Seed ferns instead were able to spread seeds through wind pollination, like modern conifers. In this hostile new world, where water was in such short supply in most of the environments, only plants that could make do with less survived. Seed ferns and conifers slowly rebounded and began to spread. Over the course of millions of years, a continent full of these plants began to slowly absorb the astronomical amount of carbon dioxide and convert it into oxygen, thus cooling the planet slightly.

Based on the lack of coal from this time period, it seems that conifers and seed ferns weren't able to attain modern tree sizes; they barely managed to get bigger than the size of large shrubs. The intense heat and arid conditions prevented them from getting bigger. Thus, carbon dioxide conversion was limited.

Even with the cooling climate, most of Pangea was still a rough place to live. The most fertile land was relegated to higher latitudes on the giant landmass. Here, ferns were able to carve out their existence. These regions of Pangea transformed into tropical environments. While the foliage and animals would be entirely different, the climate of the higher latitudes in Pangea was most comparable to modern-day Mozambique or the Dominican Republic. Small rivers and lakes formed in southern Brazil and Tanzania that could sustain life. Conifers and their palm-like relatives, known as cycads, began to congregate around these water sources, providing a small amount of shelter from the blistering heat as well as food for herbivores. As the herbivores cautiously moved to the interior, the predators followed. Thus, ecosystems began to return to the majority of Pangea.

It was here, in the interior of Pangea, that early dinosaurs would adopt an intrepid lifestyle in the background of these Pangean deserts. It may surprise you that dinosaurs did not immediately dominate the planet. But Rome was not built in a day, and neither was the empire of the dinosaurs.

Unbeknownst to the rest of the animals that scratched and clawed for supremacy, the foundation of the dinosaurs' reign was being slowly built.

The cataclysm of the end-Permian extinction did two things to lay a rock-solid foundation for the dinosaurs. First, with the end-Permian extinction, the competitive forces that kept the dinosaurs from diversifying in the first place were wiped away. Second, in the wake of all this destruction was actually unlimited opportunity. Many animals died off, which opened up many of the niches they had occupied. Even if early dinosaurs were only small generalists that ate plants and bugs, they were now an integral part of a fragile new ecosystem. There was an abundance of spaces in environments and food webs across the globe for dinosaurs to fill. With so many open niches within the environment, dinosaurs were able to rapidly evolve and diversify to fill them. These early dinosaurs didn't need to be the biggest or the most ferocious. They just needed to get a foot in the door in ecosystems across the globe. Given that all land was connected, that's exactly what they did. Soon, these small, easily overlooked animals spread all over the world. In this new world of the Mesozoic Era, the course was now set for dinosaurs to slowly rise over the rest of the animals of the prehistoric world. However, the end-Permian extinction event was just one of many climate disasters in store for the planet. For the dinosaurs' rule to come, they would still need to survive what came next.

PART 2

Late Triassic:
Journey Through the Desert

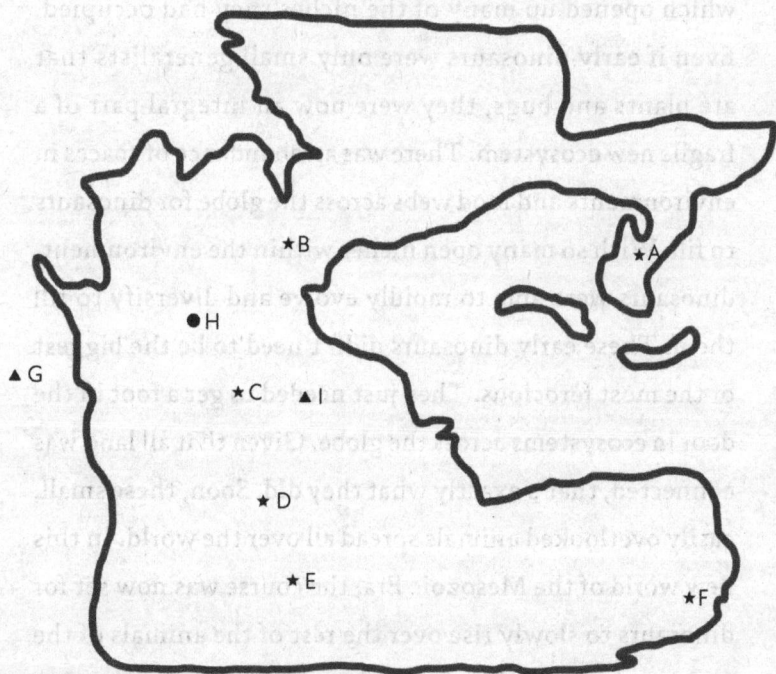

Pangea: ~237 million years ago

A - Beijing   C - New York City   E - Johannesburg   G - Wrangellia volcanism

B - London   D - Rio de Janeiro   F - Sydney   H - Manicouagan impactor

I - Central Atlantic
   magmatic province

Late Triassic Non-Dinosaurs

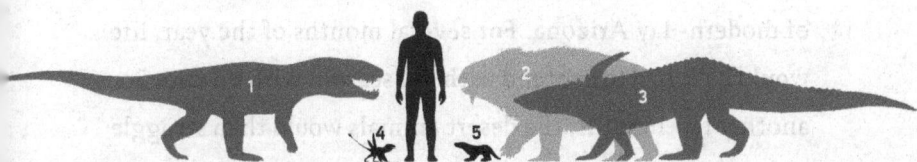

1 - *Postosuchus*, a rauisuchian
2 - *Placerias*, a dicynodont
3 - *Desmatosuchus*, an aetosaur
4 - *Peteinosaurus*, an early pterosaur
5 - *Thrinaxodon*, a cynodont

Late Triassic Dinosaurs

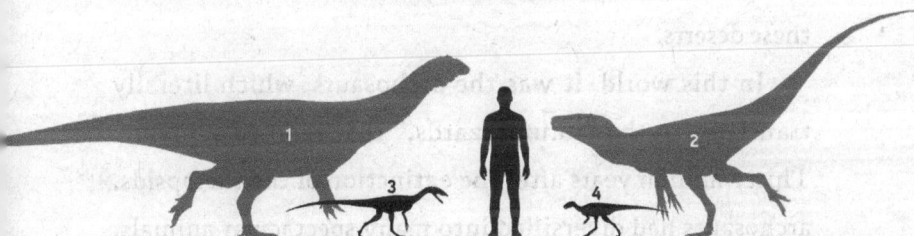

1 - *Riojasaurus*, a sauropodomorph
2 - *Herrerasaurus*, a herrerasaur
3 - *Tawa*, a theropod
4 - *Pisanosaurus*, an ornithischian

In the Late Triassic, the extremely harsh conditions of Pangea continued. Global temperatures would have been an average of 22-25 °C (71.6-77 °F), at least 7 °C (12 °F) warmer than today. While it had cooled down from the beginning of the Triassic, Pangea was still predominantly an arid desert, with little water in sight. It would have felt like living in an oven, with the sun beating down and hot dusty winds blowing past. This sweltering dry heat could be compared to the deserts of modern-day Arizona. For several months of the year, life would have had to contend with dry seasons with no rain. For another few months, the desert animals would then struggle through the wet season. Rather than relief in the form of sprinkles and showers, the massive temperature differential between the mountainous deserts and the surrounding oceans created oppressive monsoons. The harsh wasteland at the center of Pangea had only the most resilient life making up these ecosystems. Seed ferns, conifers, and cycads survived as drought-tolerant plants, and only certain archosaurs roamed these deserts.

In this world, it was the archosaurs, which literally translates to the "ruling lizards," that reigned supreme. Thirty million years after the extinction of the therapsids, archosaurs had diversified into many spectacular animals. While archosaurs existed across the globe at this time, dinosaurs specifically were still very rare and obscure animals. Many would have looked very similar to *Asilisaurus*, mentioned earlier: small and thin animals darting around and eating just about anything they could find. These early dinosaurs were generalists; many ate bugs and plants, while snatching up a fish here and there. Some walked on four legs, while other early dinosaurs experimented with walking on two legs. These nimble creatures were already relatively fast archosaurs, but those that managed to run on two legs evolved to became even faster.

At the same time, we also see the emergence of a group of archosaurs very closely related to dinosaurs, but wholly unique. Known as the pterosaurs, these close cousins of dinosaurs developed wings, making them the first flying animal besides insects to take to the skies. As insects were roughly the same size as they are today and much smaller than pterosaurs, these new flying reptiles quickly began to fill up Pangea. At this point, they had not yet evolved into iconic pterosaurs like *Pterodactylus* or *Pteranodon*, but they looked very similar, with small, lightweight bodies and large, toothy skulls. At this time, birds didn't exist yet, which gave pterosaurs full dominion over the skies.

At the pinnacle of the Late Triassic ecosystems were ancient cousins to crocodiles called the rauisuchians. They looked very similar to modern crocodiles and alligators, featuring powerful jaws and thick scales. Unlike crocodiles, species like *Postosuchus* were capable of walking, and possibly even running, on two legs. However, their hip structure was less efficient than dinosaurs', meaning they couldn't run as fast or for as long. At the top of the food chain, this was a tradeoff they could easily overcome. Another major group of archosaurs that sprang up near the end of the Triassic were the aetosaurs. These archosaurs were covered in plate-like armor on their backs, and featured skulls with an upward-curving upper lip that looked similar to the snout of a pig. Some of these herbivores, like *Desmatosuchus*, reached 4.5 to five meters (fifteen to sixteen feet) in length, and even sported huge curved shoulder spikes to defend themselves. These large lumbering archosaurs gorged on ferns close to the ground, or even used their upturned snouts to dig for roots underneath the surface. Many other archosaurs evolved tried-and-true body plans that looked remarkably similar to many other prehistoric and even modern animals. One such group were the phytosaurs, which looked nearly identical to crocodiles and gharials of today; they had long, thin snouts filled with conical teeth and armored bodies that supported legs that sprawled to the sides. Despite being very distant relatives, these predators would have behaved very similarly to crocodiles of today; hiding in large bodies of water, phytosaurs would have ambushed any animal that got too close to the shore.

While therapsids were victims of the previous mass extinction, they were not wiped out completely. The tusked dicynodonts still managed to survive. They diversified once again, and became prominent parts of the ecosystem as larger herbivores. Some, like *Lisowicia* in Poland, were similar in size to elephants. Herds of these animals, like *Placerias* in Arizona, would roam floodplains and riverbanks in search of low-lying food. A different survivor of the therapsids would actually seem familiar to us. It was small, warm-blooded, had sharp teeth, and was covered in hair. These were the cynodonts, and, while not yet considered mammals due to their primitive ear bones, they would be the ancestors of all mammals. In fact, cynodont translates to "dog teeth" because these animals had large canines near the front of the snout, and broader sharp teeth in the back of their jaws for chewing, just like our pets. While quite diverse in the Late Triassic, these small animals were at the bottom of the food chain. It would remain that way for them and their mammalian descendants for another 140 million years.

While life had seemed to settle into a time of relative peace, the climatic events of the Late Triassic would produce nearly thirty million years of unrest. Rather than the singular and horrific destruction of the Siberian Traps, the climate change of the Late Triassic was a culmination of many factors that put life through a gauntlet of trials. Part of this change began in the ocean off the west coast of Pangea, from what is roughly modern-day central Alaska down to British Columbia. In the middle of the ocean, a new province of volcanoes had sprung up at the sea floor and begun erupting, spilling lava onto the seafloor. Much like the Siberian Traps, this entire underwater region of volcanoes is estimated to have stretched out for hundreds of miles. It's estimated that this province of volcanoes released thousands of gigatons of carbon dioxide into the atmosphere. Within roughly two million years, global temperatures had spiked.

The global warming caused by this new volcanic province intensified the water cycle by evaporating more water. This obviously dries out areas and creates arid deserts. However, that isn't where the water cycle ends. All of the evaporated water can't stay floating in the atmosphere forever; it returns to Earth as rain after a certain amount of water pressure is achieved. The more intense the evaporation, the more intense the resulting rainfall will be. We see this clearly in the twenty-first century, as storms are becoming more and more intense due to global warming. The severity of tropical storms has been increasing steadily over the years. While the intensity and destruction attract our attention the most, the amount of annual rainfall increases with a warming climate, and has been over the past century. As the atmosphere warms, it can hold more moisture. Couple that with more intense evaporation caused by the increase in heat, and you get more intense and destructive monsoons. These storms cause catastrophic flooding which displaces and drowns animals, heavily erodes the landscape, and destroys habitats. This is exactly what prehistoric animals in the Late Triassic experienced as a result of this new rapid climate change.

Due to the new volcanic provinces in the Late Triassic, Earth once again spiraled into a crisis known as the Carnian pluvial episode. Not only did the world become increasingly hot, but now the residents of the massive dry deserts of Pangea were experiencing a climate they were completely unprepared for. Rain was no longer a relief from the oppressive heat; it became the new threat, as monsoons ravaged the land. Normally, modern deserts like Arizona experience only thirty centimeters (twelve inches) a year. Yet, the deserts of Pangea experienced a deluge of roughly 1.4 meters (fifty-five inches) a year. These storms didn't just happen once or twice. For roughly two million years, the wet seasons would bring with them global torrential downpours. From Argentina to Poland to even Tibet, nowhere on Earth seemed safe from the monsoons. These intense monsoons were followed by intense heat, followed again by intense monsoons, and so on, for a million years. Since the desert topsoil had been dried up, very little water could actually soak into the ground. As a result, flooding became increasingly common, as this deluge had nowhere to go except immediately downhill.

The shift was jarring for most of life, but the storms themselves did not cause any extinction. The vast majority of plants and animals did not go extinct solely because they couldn't handle the horrendous storms. In fact, this new shift to a humid environment encouraged certain life forms to prosper. The conifers, cycads, and seed ferns that survived the end-Permian extinction were much more drought-tolerant plants. Even so, they maintained a small size to conserve energy, water, and other resources. Now the high humidity and abundance of water allowed them to thrive. These plants then began to grow tremendously large; many of them became trees. The once sparsely populated deserts saw forests popping up. You might think this was wonderful for life; the planet was finally beginning to rebuild its forests. It was; carbon sequestration could improve now that carbon sinks were restoring themselves. Still, this was bad news for any animal that wasn't a dinosaur.

One of the hallmark features of a dinosaur is the hip sockets that allowed them to have an erect posture. Humans and many mammals actually have this exact same hip feature. Nearly all reptiles, like crocodiles or monitor lizards, are very low to the ground. The low posture of reptiles meant they could only eat things close to the ground. In the Early and Middle Triassic, when most plants were close to the ground, this wasn't an issue. However, in the Late Triassic, when many plants were evolving to the size of trees, any animal that couldn't reach into the trees had significantly fewer options for food. Some species couldn't handle the change in plant life, and thus died out. This also resulted in ecosystem collapse, as some carnivores that preyed upon those herbivores also died out.

Dinosaurs, on the other hand, had just struck ecological gold, as their preferred foods—conifers, cycads, and seed ferns—became even more abundant. Their erect posture made them taller than any other animal of a similar size, and thus more able to eat these new large plants. Herbivorous dinosaurs were essentially given a buffet, while many other archosaurs had their portions cut. With smaller plant- and meat-eating archosaurs dying off because of this change in plant life, the dinosaurs quickly filled their niche. This led to a sweeping replacement and diversification of dinosaurs as they transitioned from omnivore generalists to dedicated herbivores and carnivores. Not only did they become more specialized, they also became more abundant, dispersing widely. For these reasons, it's thought that the Carnian pluvial episode may have been one of the most important events in the evolution of dinosaurs. This diversification after the Carnian pluvial episode spawned several major groups of dinosaurs that engulfed ecosystems.

The first major group of dinosaurs to radiate during this time were the theropods, which eventually gave rise to iconic predators like *Allosaurus*, *Velociraptor*, and *Tyrannosaurus rex*. In the Late Triassic, these dog-sized predators, like *Eodromaeus*, would have had three sharp hand claws and a mouth filled with serrated blade-like teeth. These serrations allowed them to tear through individual muscle fibers at the microscopic level to cut through flesh like butter. Their erect posture and long legs made them much faster than other predators, positioning them as a legitimate threat in their environments and competition to other, smaller archosaur predators.

Another major group were the sauropodomorphs, who were herbivorous, omnivorous, and carnivorous as well. These dinosaurs were unique among prehistoric life at the time because many of these species had long necks. In the wake of the Carnian pluvial episode, taller conifers and cycads became more abundant in ecosystems. The herbivorous sauropodomorphs could easily reach into the trees, thanks to their increased height. Since no other animal could eat from trees because they were too short, sauropodomorphs had virtually no competition. Their sweeping success as herbivores allowed them to slowly replace many of the other large herbivores in their ecosystems, like dicynodonts and aetosaurs.

The final major group to diversify were silesaurs and their relatives, the ornithischians, also known as the bird-hipped dinosaurs. While these dinosaurs did not actually evolve into birds, they independently evolved the exact same hip structure as birds, which is a textbook example of convergent evolution. Rather than growing bigger bodies to make room for a bigger stomach (like the sauropodomorphs), this unique hip structure gave them extra space in the abdomen for bigger stomachs that could digest plants more efficiently. Even with this different hip structure, their mobility wasn't limited, meaning they were still able to outrun potential predators.

This abundance of dinosaurs didn't just carve out success in the mid-latitudes of Pangea; they began to increase their range across the equator. The increased humidity from the constant storms made nearly all of Pangea much more preferable for them. It truly is hard to overstate just how hot Pangea was. The interior desert of Pangea effectively created a heat and aridity barrier for dinosaurs, as we find dinosaurs in northern and southern Pangea before the Carnian, but not in central Pangea. With the storms of the Carnian pluvial episode, global temperatures subsided and the whole Earth shifted from extreme aridity to a brief period of global humidity. These conditions seemed to be exactly what the dinosaurs preferred, which allowed them to continue to expand their range.

Theropods and small carnivorous sauropodomorphs were the first groups of dinosaurs to cross the center of Pangea. Predatory dinosaurs in southern Pangea were finally able to migrate up to northern Pangea, and those stuck in northern Pangea were able to migrate south. This made them ubiquitous across most of the world. While they still could not compete with the large rauisuchians that dominated as the apex predators, dinosaurs began to outcompete all other smaller predators. However, herbivorous dinosaurs did not cross over into new territory. Instead, they stayed in their respective environments and continued to diversify. This wasn't because the center of Pangea was still too hot for them. In fact, the biology of large sauropodomorphs suggests they were even better adapted for life at the equator. It actually had to do with the plants that they ate. While more humid, the equator was still too hot for early conifers to grow. This meant that large sauropodomorphs and the ornithsichians stayed in their original environments. It was a time of conquest for dinosaurs as they began to cover the Earth. Even as ecosystems began to shift and change with this new influx of dinosaur diversity, there were still more climate disasters on the horizon. The next crisis in particular would not originate on Earth.

Deep from the outer reaches of space, a meteor five kilometers (three miles) in diameter hurtled toward Earth and crashed into Quebec, Canada. The impact of this meteor generated anywhere from 7.73×10^{22} to 2.30×10^{25} joules of energy. For reference, that's roughly ten million times more than is produced by one of the biggest US nuclear bombs, the B61, which generates 1×10^{15} joules of energy. While a single B61 is capable of destroying all of Manhattan, the meteor that hit Quebec left a crater over eighty kilometers (forty-nine miles) in diameter and would have destroyed hundreds of square miles. You can still see the impact crater from space, as it makes up the Manicouagan Reservoir. Anything that lived even somewhat close to the impact was instantly killed. Forests were ripped from the ground and entire ecosystems completely vaporized. Those that survived the impact may have been even worse off, as they were left in a radically different world.

A meteor impact of this scale is one of the few things that has the power to cool the entire planet. However, unlike carbon sinks, it doesn't take millions of years to cool the planet; it can happen within a few years. That's because the dust from explosions hangs around in the air. This dust blocks out the sun and can rapidly cool the area surrounding the explosion. The bigger the explosion, the more dust is thrown up into the air. The more dust in the air, the larger the area that gets blocked from the sun and cooled. It's unknown exactly how much dust was kicked into the atmosphere by the Manicouagan impactor, but it was enough to rapidly and dramatically cool global temperatures: an event known as a cold snap.

We can actually look back at recent history to see two similar events: the nuclear bombings of Hiroshima and

Nagasaki. Yes, the initial explosions reached temperatures in the hundreds of millions of degrees. However, the aftermath was also destructive. The explosions themselves ejected tons of fine particulates into the air, things like dust and ash. What's more, the mushroom clouds from these explosions were able to reach into the stratosphere for the first time, pushing the dust and ash higher up into the sky. Since they were higher up, these particulates were closer to the sun, allowing them to stay warm for longer. Heat rises while cold air sinks, so it would take even longer for these warmed particles to settle than with regular explosions, since they were closer to the sun. When enough nuclear bombs detonate and significantly block out the sun, scientists call this scenario a "nuclear winter." However, meteors are capable of producing the same effect through an "impact winter."

Thus, the unbearably hot Pangea was engulfed in soot, dust, and ash, and plunged into an impact winter. The average global temperature dropped about 2.5 °C. While this may not appear much of a difference, this change didn't happen over a million years; it happened immediately. Practically overnight, archosaurs were plunged into a long, dark winter, without any warning or time to adapt. With less sunlight available, plants were at the highest risk as they struggled to photosynthesize. Unlike in the End Permian, when plants died from the intense heat, now plants began to die off from the lack of warmth and crucial sunlight. Plants that needed more sunlight couldn't handle the change and thus went extinct. With the bottom of the food chain failing, aetosaurs and dicynodonts would have struggled to survive as well. Many species of these herbivores went extinct, yet aetosaurs and dicynodonts managed to push through. The carnivores that preyed upon these herbivores felt the ecological squeeze as well.

Archosaurs were well-suited to their hot environments before this impact winter because they were not warm-blooded animals. Many archosaurs were ectotherms, which are animals that can't generate their own body heat. Some archosaurs were also mesotherms, which are animals that could generate a tiny bit of body heat, but much less than modern-day animals like mammals and birds. Unlike humans and other mammals, nearly all archosaurs were more dependent on high air temperature to warm their bodies. In the hot and arid Pangea, the ambient temperatures were high enough to keep these reptiles alive. Without such heat, however, the internal temperature of reptiles can easily drop to dangerously low levels and they can become hypothermic. They decrease how much they move and how much they eat, things that are essential for survival. If temperatures never increase, they become more susceptible to infection, as their body slowly shuts down, and then eventually die of hypothermia. The features that made archosaurs so dominant in the Triassic quickly became an Achilles heel in an impact winter.

Dinosaurs, on the other hand, didn't seem bothered by the post-apocalyptic winter they now lived in. This isn't to say that no dinosaurs went extinct during this climate crisis.

Some species most likely did. Strangely enough, after the dust settled, dinosaurs were now becoming more diverse and successful. They were even rivaling other archosaurs for supremacy in some parts of Pangea. How on earth could this global winter not hinder the rise of the dinosaurs? If dinosaurs were archosaurs, and archosaurs were not warm-blooded, why were they not affected by the cold?

Dinosaurs were actually a unique type of archosaur that was in fact warm-blooded. More specifically, they were endotherms. This meant that they could generate their own internal body heat. The fact that they were more agile and had stronger running muscles is strong evidence of an active lifestyle typical of endothermic animals. But some of the strongest evidence comes from the way their bones grew. If you were to cut open a tree, you'd see that it has rings that are larger on the outside and smaller on the inside. These rings are formed when the tree stops growing each winter, and the spaces between rings represent active growth. The larger the space, the faster the growth. This same concept applies to bones; cold-blooded animals, like reptiles, have smaller spaces between growth rings. This means that they grow very slowly over their life. Warm-blooded animals like mammals and birds have large spaces between growth rings, meaning that they grow quickly. Despite being reptiles, dinosaur bones do not have small spaces between their growth rings. In fact, their bones seemed to grow just as fast as other warm-blooded animals.

Another piece of evidence that suggests dinosaurs were warm-blooded is that they had one of the rarest features of the time: feathers. These feathers didn't look like the types of feathers you'd see on a bird today. A feather on a modern bird is made up of a stiff bristle, called a rachis, in the middle that is covered with soft flexible parts on the outside called vanes. The feathers of the ancient silesaur were likely so primitive that they would have essentially just been the rachis of a feather. This would have looked more like the fur of animals than the feather of a bird. Since silesaurs were some of the earliest dinosaurs, and even their cousins, the pterosaurs, have been found with fossilized proto-feathers, it's extremely likely that all dinosaurs had some of these proto-feathers. They may not have been entirely covered, but they had more than other archosaurs, who had none. This is extremely significant because hair and feathers provide insulation.

All life emits heat as a byproduct. Hair and feathers keep heat from quickly escaping the body by blocking its escape path and redirecting it back toward the body, thus keeping animals warmer for longer. How or why dinosaurs developed this in the first place is still quite mysterious, especially since holding onto more body heat during the dry and arid Triassic seems counterintuitive. What this suggests is that these proto-feathers would have played a crucial role in keeping them warm during this impact winter; thus, they would not have succumbed to harsh cold like the other archosaurs and would have still been able to browse, graze, hunt, or escape.

As warm-blooded animals, dinosaurs were not at the mercy of their environments. If temperatures dropped, they would still be able to keep themselves warm. Furthermore, if temperatures were hot, as they usually were in the Triassic, they could still keep themselves cool like all other archosaurs, thanks to their air sacs. With these extra spaces on their bodies filled with cool air, they could withstand harsher heat, thus allowing them to handle a greater range of temperatures. This gave dinosaurs a huge advantage in situations where temperatures quickly changed, like an impact winter.

Yet, the impact winter wasn't the solution for dinosaurs; the removal of their competition was.

The side effects of climate change are not equal for all life. Our current climate crisis is causing many species to go extinct as the world is getting hotter and drier year after year. Yet, animals that thrive in dry, arid conditions will become more successful as time goes on. Animals like reptiles may actually prefer the world we are heading toward. On the other hand, most warm-blooded animals, like mammals and birds and especially humans, do not fare well in these types of environments. As we have seen in the end-Permian extinction, animals that can't handle the new environments go extinct; animals that can handle it take their place in the ecosystem. The archosaurs that ruled the Triassic were losing their grip in the ecosystem. In places like Poland, dinosaurs had moved into nearly all predator niches, including apex predators. The status quo had shifted.

Even after million-year-long monsoons and an asteroid impact, this was not the end of the destruction during the Late Triassic.

Underneath the center of Pangea, a new upwelling of magma pushed through to the surface to create another massive province of volcanoes. This new province is known as the Central Atlantic magmatic province, or CAMP for short, and was situated right in the middle of Pangea, where the Atlantic Ocean would eventually open up. This province did not produce one giant constant eruption of lava like the Siberian Traps. Instead, it had pulses of eruption over the course of about six hundred thousand years. You might expect the initial CAMP eruptions to do the same thing as the Siberian Traps, but the first eruptions did something different. Rather than heat the planet back up after the Manicouagan impactor, the first pulses actually cooled the planet down even further.

Jumping ahead to the eruption of Mt. Pinatubo in 1991, so many particulates of sulfur dioxide were released in the smoke that it had an effect much like that of a nuclear winter or impact winter. To give a sense of how much sulfur dioxide was released, sixteen planes were grounded in the wake of the fallout as the abundance of ash clogged up their engines mid-flight. The ash was shot out all the way into the stratosphere and created its own volcanic winter, similar to the way other large explosions can cause winters. From this single eruption, lasting twelve days, global temperatures dropped about 0.4 °C (0.7 °F).

Even more impactful was the eruption of Laki, Greenland, from June 1783 to February 1784. While temperatures were not taken back then, the climatic changes caused by the volcanic winter were well documented across the world over the course of eight months. 25% of Iceland's population died from an ensuing famine caused by the volcanic winter. Roughly twenty-three thousand British citizens died of poisoning and eight thousand also died of famine during the unexpected volcanic winter. Africa even had a significant dip in annual precipitation. Even across the Atlantic, in the newly formed United States, congressmen were caught up in unusually terrible winter storms on their way to sign the Treaty of Paris and officially end the Revolutionary War. Benjamin Franklin remarked on the serious climate change that he noticed in the aftermath:

> During several of the summer months of the year 1783, when the effect of the sun's rays to heat the earth in these northern regions should have been greater, there existed a constant fog over all Europe, and a great part of North America. This fog was of a permanent nature; it was dry, and the rays of the sun seemed to have little effect towards dissipating it, as they easily do a moist fog, arising from water. They were indeed rendered so faint in passing through it, that when collected in the focus of a burning glass they would scarce kindle brown paper. Of course, their summer effect in heating the Earth

was exceedingly diminished. Hence the surface was early frozen. Hence the first snows remained on it unmelted, and received continual additions. Hence the air was more chilled, and the winds more severely cold. Hence perhaps the winter of 1783–84 was more severe than any that had happened for many years.

However, these eruptions only lasted on the order of days or months. The pulses of the CAMP eruptions were estimated to have lasted for nearly one hundred years each. What's more, the CAMP pulsed these nearly century-long eruptions four times in the span of forty thousand years. It's thought that these first four pulses released roughly sixty billion tonnes (sixty-six billion tons) of sulfur dioxide, five hundred times the estimated amount released in Laki. Much like the impact winter caused by the Manicouagan impactor, Earth was plunged yet again into a deep and frigid winter. Yet, this winter was much more severe, as global temperatures nose-dived by at least 5 °C (9 °F). Once again, ecosystems began to collapse as plants wilted and animals starved.

Eruptions of this magnitude are not confined to the time of the dinosaurs. The eruptions of the CAMP would have actually been quite similar to what would happen if Yellowstone National Park erupted today. While these were two different types of volcanoes, both were supervolcanoes of similar size; the crater of the CAMP measured roughly ninety kilometers (fifty-six miles) in diameter, and the crater of Yellowstone is eighty-five by forty-five kilometers (fifty-three by twenty-eight miles). Geologists estimate that if Yellowstone ever erupted, it would cover the entire US in ash. At the epicenter of the eruption within the national park, the area would be submerged in over five meters (sixteen feet) of ashfall. All the surrounding states could experience anywhere from one to five meters (three to sixteen feet) of ashfall. Cities would be crushed underneath the downpour. From coast to coast, Americans could face anywhere from one to forty centimeters (0.4 to fifteen inches) of ashfall depending on how close they were. Even Mexico and Canada would receive up to a centimeter (0.4 inches) of ashfall. The eruption itself would practically destroy the nation.

Any survivors in North America, and indeed the rest of the world, would then have to contend with the ensuing volcanic winter. If the eruption of Laki was capable of causing famine across Europe and Greenland, it's not hard to imagine that the eruption of Yellowstone would cause famine on a global scale, as temperatures plummeted and the sky darkened. To say that we would be knocked back to the Stone Age wouldn't even be accurate, as our Stone Age ancestors could at least hunt and gather. The new freezing global temperatures would mean there would barely be any plants to gather and even fewer animals to hunt. It could very well cause one of the worst mass extinctions in Earth's history, and possibly the end of humanity.

Things that can quickly cool the globe are not a solution to global warming. As we can see, while the warming of the Late Triassic was halted by these rapid cooling events, it came at the expense of life and the extinction of many species. Unlike the slow and gradual destruction of global warming, these global cooling events were short and violent. They took place at a rapid pace, which is a massive problem for ecosystems. Unfortunately, evolution doesn't work on the scale of centuries, so only life that is already adapted to dark, cold, and barren landscapes is primed to survive. How many animals do we know that can handle such a change? Not many. More importantly, humans can't either. This doesn't mean that global warming isn't a serious issue. Our current crisis is still taking place faster than life can adapt. It just means that many things that can cause global cooling, like meteor impacts, volcanism, or even the explosions of large-scale war, are just as serious threats as carbon emissions in an equal yet opposite way.

The grand finale to the cataclysms of the Late Triassic was the last and biggest pulse of this volcanic province. Rather than the short and violent eruptions that preceded it, this eruption was longer and slower. The final eruption was not notable for how much sulfur dioxide it released, but instead for emitting literally tons of carbon dioxide in less than a million years. With this last eruption, the CAMP released roughly eighty thousand gigatons of carbon dioxide. This is roughly two thousand years of carbon emissions at our current rate. This massive amount of carbon dioxide not only stopped the volcanic winter caused by the previous eruptions; it warmed the planet back up to how it was before the CAMP began erupting. You might think this negated all of the damage already done, but that's unfortunately not how climate change works. All it did was make life harder for those who managed to survive the volcanic winter; those species that managed to stay alive during the harsh winters now dealt with brutal heat. No one could win.

In the end, the CAMP eruptions constituted one of the five deadliest events in Earth's history. In the span of thirty million years, life was hit with four life-changing events that reshaped the planet. Most of the archosaurs that dominated the Triassic went extinct; the armored aetosaurs, aquatic phytosaurs, and the predatory rauisuchians all succumbed to the destruction of the CAMP. Only a small group of archosaurs that would eventually evolve into crocodiles and alligators survived. Even the dicynodonts were eradicated completely, with only a few species of cynodonts surviving to give rise to the mammals, although the cynodonts were so crippled that their mammalian descendants were forced to carve out life at the bottom of the food chain for the next 140 million years.

Practically all that was left were the dinosaurs and their cousins, the pterosaurs. While it's still quite mysterious why these dinosaurs went extinct and all the others didn't, it may be due to their early success. As the apex predators of their environments, they may have become too reliant on other Triassic fauna as prey; once these foundational prey went extinct, they too died off. However, for the rest of the dinosaurs, their unique combination of metabolic features and highly efficient ways of moving made them capable of surviving just about any type of global change. In a similar way to the end-Permian extinction, dinosaurs were still not massive or specialized in their diets or niches; they were still flexible enough to weather ecological changes. Or at least, they were more flexible than the other archosaurs, who were bigger and more specialized than them.

It is also important to note that dinosaurs were most abundant in places like Argentina and South Africa. Some were also becoming common in European countries like Poland and Germany. What all of these countries have in common is that they are all coincidentally outside of the extent of the lava flows of the CAMP. While it's impossible to know for sure, it's likely they were also far enough away to receive relatively little ashfall compared to any other archosaurs or dinosaurs that lived closer to the equator. Most dinosaurs may have survived simply because they were in the right place at the right time. While evolution is driven almost entirely by the biology of an organism and how it relates to its environment, luck plays at least some part in evolution as well. This selective force, known as genetic drift, is when a crisis strikes and those who survive and die off are technically random, even if the victims may have better genes. A bug may be perfectly adapted for its environment and niche, but it doesn't matter if a person happens to step on it. Likewise, archosaurs who lived closer to the CAMP eruptions died off while the dinosaurs who were lucky enough to be farther from these eruptions lived on.

For nearly fifty million years, during the entire Triassic Period, dinosaurs were the ultimate underdog, and it was finally paying off. Despite suffering through the Siberian Traps, global monsoons, meteor impacts, and even more volcanism, these animals not only endured; they went on to thrive.

PART 3

Early Jurassic: *Manifest Destiny*

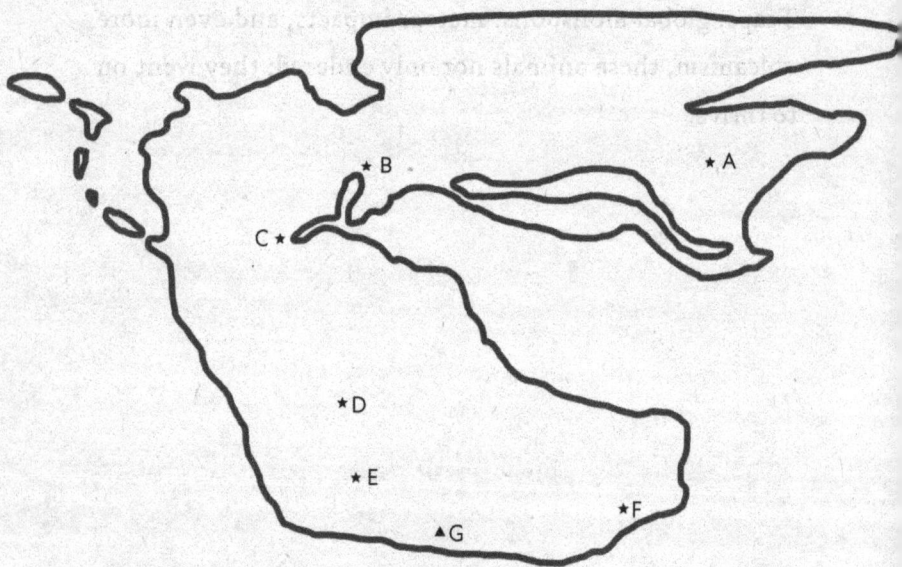

Pangea: ~201.3 million years ago

A - Beijing	C - New York City	E - Johannesburg	G - Jenkyns Event volcanism
B - London	D - Rio de Janeiro	F - Sydney	

With the end of the Triassic Period 201 million years ago, Earth entered the Jurassic Period, the second of three time spans that make up the Mesozoic Era. During the Early

Early Jurassic Dinosaurs

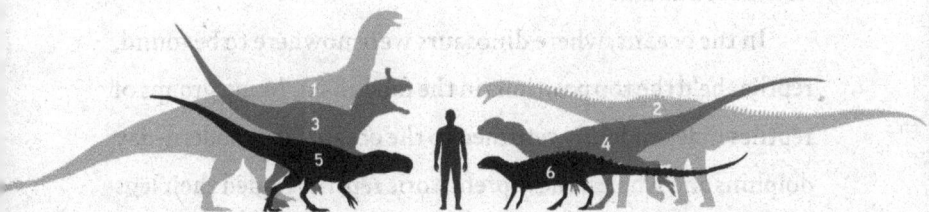

1 - *Lishulong*, a sauropodomorph
2 - *Vulcanodon*, an early sauropod
3 - *Cryolophosaurus*, a polar theropod
4 - *Dilophosaurus*, a theropod
5 - *Monolophosaurus*, an early tetanuran
6 - *Scelidosaurus*, a thyreophoran

Jurassic, and in the wake of the end-Triassic extinction, life had shifted radically. Dinosaurs had filled just about every major niche in their environments, from apex predators to megaherbivores exceeding one tonne (1.1 tons) in size down to smaller bug-eating predators, known as insectivores, and tiny herbivores. At this point in history, it can be easy to think that the environment was exclusively filled with dinosaurs from all of the mass extinctions and ecological shake-ups. While it's true that dinosaurs were the dominant group of animals at the start of the Jurassic Period, Earth was teeming with many other amazing creatures. Amphibians, lizards, and turtles still made up much of the ecosystem. Although archosaurs had been brought to their knees by the end of the Triassic, the ancestors of crocodiles and alligators were still around. Many species, like *Protosuchus*, carved out a meek existence among the dinosaurs as smaller secondary predators in the food chain.

In the oceans, where dinosaurs were nowhere to be found, reptiles held the top positions in the food chain. Many groups of reptiles independently returned to the ocean. Like modern-day dolphins and whales, these prehistoric reptiles traded their legs for flippers. These oceans were teeming with tons of unique and strange animals. Long-necked plesiosaurs ambushed fish, while the speedy and powerful pliosaurs swam after larger prey. Some even looked eerily similar to animals we see today. Ichthyosaurs adopted a body reminiscent of dolphins, with long snouts and dorsal fins. Ancestors of crocodiles returned to the seas as the thalattosuchians—keeping their iconic scaly exterior and toothy snout, but sporting short, stumpy flippers. The ocean became such a diverse and deadly place that not even dinosaurs would dare venture into the waters.

In the skies, the pterosaurs had their breakout. The ecological space was theirs for the taking, and pterosaurs were found in the skies throughout the whole world. Pterosaurs like *Dimorphodon* were roughly the size of a bull terrier and featured a disproportionately large head with a set of snaggle teeth. These fliers soared across Pangea in search of insects or small lizards.

It was also at this time that the ancestors of modern mammals evolved from the cynodonts. These mammals would have looked like shrew- or chipmunk-like animals skittering around in the underbrush. Because so many of the niches across the globe had been taken by dinosaurs, these early mammals would have remained small insectivores that burrowed to hide from the lethally quick dinosaurs. Additionally, like echidnas and platypuses, they would have still laid eggs in these burrows. Mammals would more or less stay like this for the entire reign of the dinosaurs.

At this point, the eruptions of the CAMP had ceased and carbon dioxide was finally on the decline. Carbon dioxide was decreasing at a very slow rate, suggesting that carbon sinks were doing their job; plants were absorbing carbon dioxide and releasing oxygen in return. This time period saw an increase in oxygen, something that hadn't been seen for millions of years. It's estimated that oxygen levels were much higher than they were in the Triassic, and that they were similar to, if not slightly higher than, what we experience today. Higher oxygen levels are often correlated with cooler temperatures. However, carbon dioxide is the main driving force when it comes to our climate. As mentioned earlier, carbon dioxide is considered a greenhouse gas, which are molecules that absorb additional heat, reemit this heat back to the Earth, and warm the planet. Yet, oxygen is not necessarily an "anti-greenhouse gas." Plants absorb carbon dioxide during photosynthesis and create oxygen as a byproduct. As plants absorb carbon dioxide, they draw from the total amount of greenhouse gases in the atmosphere. Bigger forests draw more carbon dioxide, which leads to less greenhouse gases in the atmosphere warming the planet. and more oxygen. Thus, high oxygen levels are often related to a decrease in temperature.

This produced a long cool interval during the Early Jurassic. Global average temperatures reached down to about 19 °C (66.2 °F) at its coldest point, yet this is still slightly hotter than today. The poles had no permanent ice caps during the Early Jurassic like they do today; it was cool enough that it could snow in the winter, but the temperatures were moderate enough to support life throughout the globe, including the equator and the poles. While the dinosaurs were transitioning out of the intense heat and dry Pangean deserts, they still lived in a warmer world than our current one. While dinosaurs did live in climates like ours, and even climates colder than ours, much of the Mesozoic was warmer than today.

Earth has gone through many different climate states before, during, and after the reign of the dinosaurs. A coolhouse climate with global temperatures that are roughly 18-22 °C (64.4-71.6 °F), similar to our modern climate, only encompasses 18% of the past five hundred million years. Ice ages encompass 13%, even more rare. That leaves the vast majority of ancient history hotter than today. Thus, Earth is usually a much warmer planet than it is today. So, if Earth has a track record of usually being warmer than today, what does this say about our modern world? Is global warming inevitable? Are we currently set on a course that can't be stopped? Without any human intervention, global warming as a whole is *probably* inevitable. Just from simple probability, based on those percentages, it's very likely that our planet will get warmer in the future. But that's okay. Even in a world much warmer than the one we currently live in, many organisms will be able to survive. The reign of the archosaurs and the rise of the dinosaurs show that many animals are capable of thriving in hotter and drier climates. It just means life will look much different in the future. However, what isn't inevitable and does require our attention is how fast our planet is warming.

The climate disasters that the dinosaurs survived thus far happened in about a million years or less. The global warming caused by the final CAMP pulse, for example, took place in less than a million years. Yet this cool interval of the Early Jurassic, which saw a total drop in temperature of roughly 5 °C (9 °F), took place over the course of twenty million years. What life struggles to adapt to is rapid change in global climates, not necessarily the direction or even the magnitude of climate change. Evolution takes time. The magnitude of change in the Early Jurassic would have constituted another climate crisis if it happened at a tenth or even a hundredth of the rate. Instead, it produced a stable climate for dinosaurs. Thus, climate change, in and of itself, is neither good nor bad; it just is. The rate at which we are releasing carbon dioxide is comparable to many of the eruptions of the volcanic provinces that brought life to its knees. However, we do have the ability to change it. No one could stop the disasters of prehistory from striking, yet humanity can stop the disasters of today; you and I can.

During the Early Jurassic cool interval, plants were diversifying and forests were growing. Earth was no longer a giant hot and arid desert. It became much more hospitable to plant life. The equator, which had been so hot and dry that plants could barely grow there, had finally cooled down. Plants could now inhabit it if they were sufficiently adapted to the heat. What's more, forests composed of these plants could spread across the equator into the other parts of Pangea. As plants began to spread across the globe, they began to evolve differently at the different latitudes. At the equator, small-leafed conifers and cycads adapted to dry deserts began popping up. In the mid-latitudes, forests sprang up filled with ferns and horsetails, with cycads, seed ferns, and conifers all growing into massive trees for the first time. Even at the poles, large-leafed conifers and gingkos filled the area. It's hard to imagine places like Antarctica as anything but an expansive white landscape, but in the Early Jurassic, it would have been exceedingly lush in comparison to today. Rather than an icy tundra, it would have looked very similar to places like the North Island of New Zealand. With an abundance of plants everywhere, including extreme environments like the equator and the poles, all of Pangea was suitable for all types of life. The varied plant life across Pangea encouraged animals to evolve into different forms to fit the unique niches around the world.

Thus, in concert with the cooling of the Early Jurassic and the diversification of plants, dinosaurs evolved rapidly, on a scale never seen before. The herbivorous dinosaurs, who earlier could not migrate across Pangea because plants were limited to the mid-latitudes, could now cross the equator. The large sauropodomorphs and smaller ornithischians began to move north into North America and Europe. Moreover, it wasn't just herbivores that began to spread out; the predatory theropods began to fill Pangea as well in search of new prey. The interconnectedness of Pangea and the very few mountain ranges allowed dinosaurs to migrate out across the entire globe.

Despite the entire planet getting cooler, the poles were still warm enough for dinosaurs to live there. In fact, within two million years after the end-Triassic extinction, the poles had robust ecosystems filled with an assortment of unique dinosaurs. Near the North Pole in what is now China, the elephant-sized sauropodomorph *Lishulong* would have grazed among the trees with its long neck. Near the South Pole, in what is now Antarctica, *Cryolophosaurus*, a six-meter-long (twenty-foot-long) predatory theropod featuring a unique pompadour-shaped crest, prowled the forests for prey. With dinosaurs ranging from the poles to the equators, they could now be found in every single environment on Earth.

As the dinosaurs spread out across Pangea, they went through a biological process known as dispersal. As a population of animals grows, they begin to spread out from the epicenter of where their population is located, sometimes even crossing over barriers like large bodies of water, mountain ranges, or swaths of forests. When they cross over into this new land and are no longer connected to the main population, they begin to reproduce in isolation. This isolation, over a long period of time, eventually leads to the creation of new species, as the animals slowly become less like the original population they came from. If a group of birds flock to an island, they have dispersed away from their main population. If they never flock back, they will never be reintroduced to the genes of the original population, and thus will slowly change their features as they adapt to their new environment. At first, these will be small changes, like the shape of their beaks, the color of their feathers, or minor size changes. Over millions of years, though, this could create an island of birds radically different from the original population. This process of new species originating, known as speciation, happened for dinosaurs during the Early Jurassic, but on a global scale, helping turn a generic group of animals into a whole menagerie of unique creatures.

During this global dispersal, at the end of the Triassic, the three major groups of dinosaurs at the time (theropods, sauropods, and ornithischians) began to split into new and different groups. The theropods began to split into two unique forms. On one side were the ceratosaurs. These predatory dinosaurs evolved more robust skulls and shortened arms; in fact, their arms would become even smaller and more useless than those of T. rex. On the other side were the tetanurans. These predators prioritized stiff tail muscles and more advanced air sacs over stronger skulls. With better tail muscles and lungs, they were better suited for chasing down prey.

Sauropodomorphs continued to grow in size, and an offshoot became so big that they transitioned to walking on four legs. These became proper sauropods. On four legs, they could support a more massive body on pillar-like limbs. By also developing hollow bones, like those of birds, they could avoid being crushed by their own weight. Add to this a longer neck, and they could eat more food. Like a giraffe, they could more easily reach up into the treetops for food and use that long neck to help them cover large swaths of ground without moving. Movement takes energy, and to move such a massive body would require an immense amount of energy. Therefore, to save energy, sauropods could stay in one spot and move their necks across the ground and into the trees to graze.

The ornithischians split as well, into two distinct groups. The first branch, known as the thyreophorans, developed armor over their body. Like crocs and alligators, they had bony patches known as osteoderms that covered their backs and sides. The other group was the neornithischians, or "new ornithischians." The reason for their name is that they all had a thickened enamel on the inside of their teeth that allowed them to do something entirely novel in dinosaurs: They could actually chew their food. We might take this for granted, as animals that naturally chew their food, but no other dinosaur prior to this did so. When you watch a bird eat, it swallows its food whole, no matter how big. The reason we chew our food is to break it down and allow for quicker digestion in the stomach. Plants are especially tough to break down compared to meat. Animals who don't chew their food have usually developed several different techniques to digest it. Some herbivores, like a few species of modern birds, used gastroliths: small stones that were swallowed and used to mash up their food. Other herbivores employ something called gut fermentation, where gut bacteria play a more active role in further breaking down plant material. A completely unique way that humans got around this was by cooking our food, which also breaks down complex molecules for easy digestion. Regardless of what method an animal evolves, it will take much longer to digest food and access the nutrients if it is not broken down beforehand.

As oxygen increases during this cool interval, we also see dinosaurs become much larger. Ornithischians diversified into much larger dinosaurs. Some of their descendants, like the early armored dinosaur *Scelidosaurus*, reached up to 3.8 meters (12.5 feet). Sauropodomorphs were being replaced by the iconic sauropods. Sauropods like *Vulcanodon* now reached up to eleven meters (thirty-six feet). Even the predatory theropods were getting bigger and more fearsome. Predators like *Dilophosaurus*, which dwarfed polar bears in size, ruled as apex predators. Dinosaurs had now become the biggest animals to ever walk the Earth up to that point, dwarfing those of the Triassic and Permian (and these weren't even the biggest dinosaurs to ever exist; they got *much* bigger later on).

Oxygen has commonly been related to the size of an organism. As oxygen increases, life usually gets bigger. This is true for both plants and animals. For example, during the

Carboniferous Period, oxygen levels were at their highest in Earth's history. As a result, life attained some of its biggest sizes yet. The emergence of the first trees began during this time. Primitive trees covered in scaly bark, known as *Lepidodendron*, reached up to fifty meters (160 feet) and congregated into some of the first forests. Insects reigned as they reached body sizes you would expect in horror movies. Dragonflies, like *Meganeura*, with over seventy-centimeter (two-foot) wingspans; centipedes, like *Arthropleura*, reaching 2.5 meters (eight feet) in length; and nearly fifty-four-centimeter-long (two-foot-long) arachnids called *Megarachne* are just a few of the giant insects that stalked the swamps of this ancient time. Even after the rule of the insects during the Carboniferous Period, the vertebrates of the Permian grew to some of the biggest sizes seen thus far in Earth's history. Oxygen had declined slightly during the Permian, but it was still relatively high compared with all of Earth's history. While animals like *Inostrancevia* and *Scutosaurus* were not impressive compared to dinosaurs, these were giants compared to the other prehistoric animals of the time.

This begs the question: Was the increased size of dinosaurs due to more oxygen in the atmosphere? Not exactly. While it's unknown exactly how much oxygen was in the atmosphere during the Early Jurassic, many think that it was comparable to today, rather than the abundance seen in the Carboniferous or even the Permian. But today, we don't have gigantic mammals roaming the Earth. So, why did dinosaurs get so much bigger than the mammals of today in such a short time?

Research suggests that it goes back to how the dinosaurs breathed. Remember, dinosaurs had unidirectional lungs with multiple air sacs along the pathway of their lungs; this unique respiratory system made them, and all archosaurs, extremely efficient breathers. This came in handy at the end of the Permian, and throughout the entire Triassic, because oxygen was in very limited supply. Hence the ancestors of mammals, like therapsids, dicynodonts, and cynodonts, either went extinct or became much less diverse during this time. As oxygen within the atmosphere began to increase, the efficient lungs of dinosaurs were already primed to take advantage of the new abundance of oxygen. Having more oxygen held in the body due to air sacs meant that dinosaurs could provide more energy to their muscles. With more energy, muscles could get bigger and stronger. They could support a more active lifestyle, or bigger bodies. Not only did the abundance of plant life and stable climate benefit the dinosaurs, but the extra oxygen took their highly efficient bodies to the next level. It was this perfect storm of environmental and climate characteristics that helped them become exceedingly successful as time went on.

However, while the Early Jurassic saw a long stretch of relatively stable climate, this stability ended around 183 million years ago. In Pangea's South Pole, another province of volcanoes sprang up, spanning from what is now Antarctica to South Africa. This province of volcanoes, known as the Jenkyns Event, flooded southern Pangea in roughly three million kilometers2 (one million miles2) of lava. For reference, this would be like a third of the continental US being covered in lava. As with many volcanic provinces before, these volcanoes released enough carbon dioxide to raise the temperature of Earth by as much as 8 °C. All of the cooling that had taken place during the last twenty million years was abruptly undone by a couple hundred thousand years of intense heat.

This heating was most significant at the equator, and even affected the Hadley cell, which is a region of the atmosphere that circulates warm tropical air to the mid-latitudes and colder air back to the equator. Today, the Hadley cell is important in shaping regional weather, and even tropical storms. In the Early Jurassic, this intense heating caused more storms and even created more mega-monsoons near the equator. This intensification of the water cycle also increased evaporation at the equator, creating a hot and dry desert belt in the center of Pangea once again. While the equator experienced new changes, the vast majority of Pangea was still covered in exceedingly lush and cool forests. What would normally be a global disaster was actually reduced to regional climate changes. Even more peculiar was the effect this had on dinosaurs as a whole.

The sauropods took advantage of global warming. While their cousins, the two-legged sauropodomorphs, fizzled into extinction, the massive sauropods grew to gigantic sizes, with even longer necks. They became the biggest land animals ever to exist up to that point in Earth's history.

Sauropods were able to leverage their special physiology to become the dominant herbivores of this period. Unlike most dinosaurs, sauropods had a strange metabolism that we don't see in many other dinosaurs. By all measures, they appeared warm-blooded: They grew faster than any other land animal ever. They had hollow bones, like birds, keeping them lightweight. Together with their dinosaurian features, like air sacs and erect posture, it suggests they were active to some extent. The vast majority of these dinosaurs lived in hot environments; many of them actually preferred the hot and dry equator that formed from this climate change over the cooler and wetter environments farther from the equator. Thus, it seems sauropods may not have been warm-blooded in the way we normally think, but were instead a mixture of endothermic and poikilothermic.

For most animals, there is a line distinguishing between the warm-blooded and cold-blooded. But the metabolism of some animals doesn't fit neatly into this binary placement, especially those in our ancient past when they are changing over time to fit into an environment. Poikilotherms are animals that can't maintain a stable internal temperature. Their body temperature may rise and fall very easily as a result of the temperature of their environments. This might sound very foreign to us because we, as humans, have the total opposite type of metabolism. We are homeotherms, meaning we have the same body temperature at nearly all times, regardless of whether it's hot or cold outside. However, lizards basking in the sun need the sunlight to warm themselves up.

Most warm-blooded animals today are both endothermic and homeothermic—meaning they can generate their own body heat (endothermic) and keep a stable temperature (homeothermic)—whereas most cold-blooded animals cannot generate their own body heat (ectothermic), and are dependent on environmental temperatures to warm them (poikilothermic). Sauropods were able to generate their own body heat (endothermic). Yet, in order to keep their unusually gigantic bodies active, they lived in warmer environments to further prevent their metabolism from declining.

While the sauropods secured the role of the dominant megaherbivores, ornithischians flourished at all the other size classes. Sauropods preferred much hotter environments to attain their massive sizes, which left many niches open in relatively colder environments. The farther from the equator, the less diverse sauropods became, and therefore the more open herbivore niches there were. This is where the ornithischians capitalized. This isn't to say that ornithischians didn't live in warmer climates closer to the equator. Near the equator, they filled whatever niches were open as medium to small herbivores. Yet, as the environments got colder, ornithischians made up more of the fauna. Evidence suggests that ornithischians were both endothermic and homeothermic, unlike sauropods. While they were unable to attain the immense sizes of the sauropods, they were less reliant on the temperature of their environment for their success. Therefore, they were actually better adapted for living at mid to high latitudes. It was in these cooler environments that ornithischians were at their most diverse. As for the neornithischians, who could chew their food, this meant they could eat tougher plant material that other non-chewing herbivores were avoiding. While sauropods and thyreophorans sought out easily digestible plants, neornithischians ate tougher plants. This complete lack of competition allowed them to explode in diversity.

However, at a certain point, an herbivore can become so big that it is nearly impossible for an individual predator to hunt it. Predators usually hunt prey that is less risky to catch. A large and healthy individual may provide more food if caught, but its health, strength, and size make it harder to catch, and even dangerous if it fights back. Prey animals that are smaller, slower, and/or weaker are often easier to catch, though they provide less food for the predator. The size difference between most sauropods and theropods had become so large that a fully grown sauropod was a risky option for a single predator. Ornithischians, on the other hand, couldn't reach these massive sizes and needed different defenses. Armored dinosaurs answered this by growing bigger and more complex defenses. Rather than simply having tough bumpy skin like that of an alligator, they developed actual defensive structures. Spikes, plates, and thick osteoderms adorned their bodies and turned them into walking tanks. In addition, spikes and clubs grew from the tails of many species, protecting them even further. Neornithischians had neither size nor intimidating defenses. Therefore, many of them prioritized the development of strong leg muscles or maintained small sizes, all with the goal of being as quick as possible.

Theropods became exceedingly diverse as well. While theropods were already prominent in the Early Jurassic as the only predatory dinosaurs, the ceratosaurs and tetanurans all looked very similar. These predators then began to radically shift into unique bodies as the unique global environments encouraged innovation. Some of them grew exceedingly big. Their jaws became bigger and stronger to deliver even deadlier bites; their arms and claws grew into more fearsome weapons. Massive predators like *Dilophosaurus* were replaced by even bigger ceratosaurs and newer breeds of tetanurans that easily surpassed the biggest land predators of today. These large predators seemed to prefer the hotter dry habitats across the globe, with a few exceptions.

It didn't always pay off to be the biggest or the scariest. Some tetanuran theropods evolved down a different path by keeping a modest size, or even shrinking. These smaller theropods developed long legs for running, and some even developed longer arms that may have helped in grasping prey. While it's thought that most theropods had some type of hair or early feathers, these smaller species were undoubtedly covered from head to tail in early feathers. Their speedy features, insulation, and high metabolism were all hallmark features of warm-blooded animals. These theropods were the coelurosaurs, and they were extremely diverse. Their high metabolism helped them inhabit all sorts of environments. It's through these dinosaurs that we will see one of the most interesting evolutionary stories in the Mesozoic.

Now, at the end of the Early Jurassic, dinosaurs had finally become the dominant animals on Earth. They diversified from the hothouse climate of the Triassic to the cool interval of the Early Jurassic, through global warming brought on by volcanism. They were hardly the same group of animals. As the Early Jurassic began to close, some actually preferred the hot and dry equatorial climates, while others preferred the cooler and/or more humid environments in the mid to high latitudes. This radical change not only established them as a global group of animals, but also prepared them for all the other shifts and changes that would soon take place throughout the rest of the Mesozoic.

As evidenced by the Triassic, life is fragile, and can be extinguished easily on a planet that is constantly changing. However, it is also extremely resilient and adaptive, as evidenced by the introduction of the Jurassic. Life refuses to die. For over half a billion years, complex life has done whatever it could to continue its existence. It moves to foreign and strange lands to find more opportunities. It shapes its body to deal with increasingly difficult challenges in an ecosystem. It restructures its physiology to survive the changes of a new climate. In the case of the dinosaurs, when a combination of adaptations that have accumulated over millions of years (like their specialized lungs and hyper-efficient legs and hips) meets a new opportunity within their environment (like the introduction of a cooler and more humid climate), they take advantage of that change to secure their survival.

Dinosaurs are a great example of a group of animals whose global success was closely tied to climate. Yet, we so often forget that we are animals too; we have an environment and niche where we are the fittest. Yes, the advancement of science, technology, and medicine have altered our fitness and survival beyond what most animals are capable of. However, we know that the path we are headed on is not the best climate for us. A warmer and drier planet is not only not conducive to our biological needs, but our long-term success also depends on societal needs that are affected by global warming. Just as climate shaped the evolution and rise of dinosaurs, climate will shape the future success of humanity. Do you know what that future should look like?

The climate debate seems increasingly complex because there are so many problems. How hot is too hot? How fast should warming be? How quickly can we reverse warming? Should we focus on investing in renewable energy? Should we focus on reducing carbon emissions? Can we do both? Can we afford both, in terms of time and money? What about deforestation? How do we protect and conserve our current ecosystems?

It can feel overwhelming. Like a string knotted many times, it's hard to even know where to start untangling a problem this complex. Therefore, one approach that might clarify how we effectively resolve our climate crisis is to take note of the dinosaurs. Certain species fit best into certain environments.

Even though dinosaurs first evolved during a time that was on average hot and dry, they were secluded to the cooler southern half of Pangea during the Triassic. As these climatic conditions spread across Pangea during the Early Jurassic, we see the range of the dinosaurs expand as well. Could we not think similarly for humans and the multitude of other species that evolved alongside us? Just as we can look at the environments dinosaurs prefer to live in, we can look at the environments that humans prefer to live in.

To know what the best climate is for human survival, we can look at where humans prefer to live. Even though humans occupy almost every corner of the planet, we're surprisingly picky in where we live. Most of humanity is concentrated in temperate areas that are on average 11–15 °C (51–59 °F). The eastern half of the US, Western Europe, the Mediterranean, Central Asia, eastern and coastal China, Japan, South Africa, and central Argentina: These places fit into a highly specific climatic niche of humans. Not only are temperate environments where we see the highest density in human populations overall, it's also where we see the highest concentration of crops and livestock. It's even where we see the highest density of global GDP. This isn't just true for today, or the past couple decades, or even the past few hundred years. For over six thousand years (which is about as far back as we can reliably tell), humans have lived in temperate environments.

However, if we continue with our current trend of global warming, temperate forests will be significantly affected. Climate modeling shows that as temperatures increase, temperate forests will recede closer to the poles and away from the major cities and cultural zones they currently occupy; instead, they will move toward areas that are currently less inhabited. Similarly, the equator will become increasingly arid and dry, creating an expansive desert. Extremely arid deserts only make up about 0.8% of all land today, but that is estimated to increase to up to 19% of Earth's surface by 2070. It's estimated that, in the next fifty years, most of the global population will no longer live in temperate climates. Most people will be living in climates we're not meant to live in. An estimated 3.5 billion people will be living in deserts. Crops will be harder to grow. Livestock will struggle to live. If nothing else, global macroeconomics will radically change. Life will become harder for the majority of humanity, even those that live in relatively wealthy countries. Therefore, conserving and rebuilding temperate environments on Earth can be an actual end point climate remediation should strive for.

From this we need to remember that, in our current climate crisis, life does not give up easily. While some species will go extinct and some ecosystems may collapse, life as a whole will still persist into the future. As we work to counteract the rapid heating of our planet, life is also working to prepare itself for whatever tomorrow brings, whether it's a warmer day or a cooler day. There has yet to be a climate disaster so damaging that it caused the complete extinction of life. Then again, it only takes one event that is destructive enough to eradicate life forever.

PART 4

Mid-to-Late Jurassic:
Empires Arise

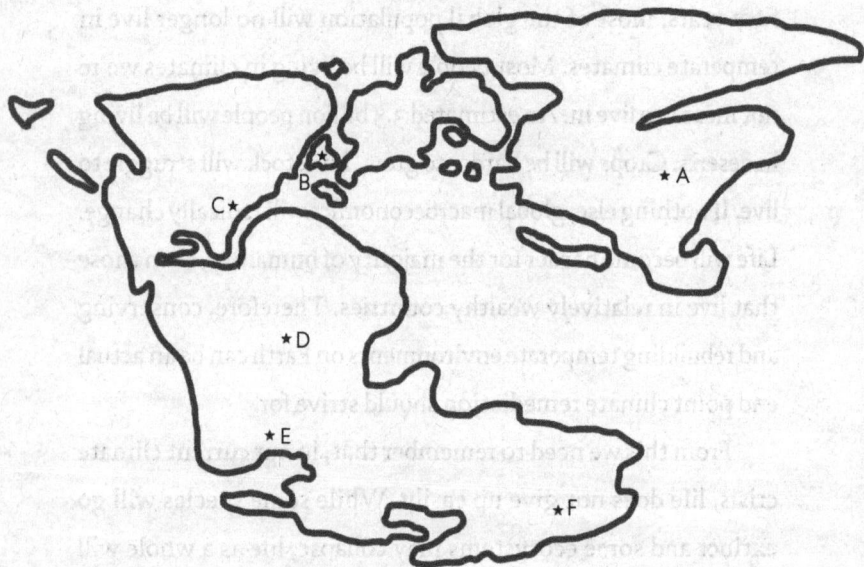

Pangea: ~174.7 million years ago

A - Beijing   C - New York City   E - Johannesburg
B - London   D - Rio de Janeiro   F - Sydney

Earth began to look very different during the start of the Middle Jurassic period. Pangea, which had been the only landmass for the past seventy-six million years, was slowly

Mid-to-Late Jurassic Dinosaurs

1 - *Apatosaurus*, a diplodocid
2 - *Mamenchisaurus sinocanadorum*, the largest mamenchisaur species
3 - *Cetiosaurus*, a eusauropod
4 - *Allosaurus anax*, the largest allosaur species
5 - *Megalosaurus*, a megalosaur and the first dinosaur ever discovered
6 - *Yangchuanosaurus*, a metriacanthosaur

being ripped apart by the tectonic forces that brought it together; in fact, the breakup of Pangea happened at about the same pace at which your fingernails grow. Near ground zero of the Late Triassic CAMP eruptions, central Pangea split apart. Like two conveyor belts moving in opposite directions, the thick continental crust was pushed outward and replaced with thinner yet denser crust. This new thinner crust created a lower elevation. Naturally, water from the surrounding ocean then filled it in to create a newer and smaller ocean. While nothing like it is today, this new ocean would grow and become the Atlantic and Indian Oceans.

1 - *Stegosaurus*, the largest stegosaur
2 - *Ceratosaurus*, a ceratosaur
3 - *Camptosaurus*, an ornithopod
4 - *Proceratosaurus*, a coelurosaur
5 - *Enigmacursor*, a neornithischian
6 - *Archaeopteryx*, the first bird

Antarctica. While Europe and Asia were still connected, the higher sea level of the time caused a seaway, called the Uralian Sea, to cut across these two regions.

The breakup of Pangea didn't stop with just Laurasia and Gondwana. As we can see from our modern globe, there was still much more rifting of the continents to come. The Indian Ocean grew for another twenty million years, until the end of the Jurassic, with the Atlantic slightly growing as well. The massive separation in the Indian Ocean eventually led to Madagascar, India, Australia, and Antarctica becoming their own distinct continents, apart from the combination of South America and Africa, known as Samafrica. Yet, at this point, what would become the southernmost regions of South

America and Africa were still connected to this polar continent by a small land bridge.

The breakup of Pangea and the creation of the new oceans had a profound effect on the climate of the Jurassic. As these early oceans began to snake through the freshly-split continents, ocean currents were redirected between the continents. These new oceans exposed more land to cooler ocean air and humidity. The introduction of these ocean currents ended the overall intense heat that was so common in Pangea. These cool currents also reduced the intense seasonality of the climate. The cycle of intense dry seasons and mega-monsoons that plagued Pangea began to fade away. Yet, unlike in the Early Jurassic, when the climate formed a gradient from the equator to the poles, unique climates now began to pop up all over the world.

One such unique region was Europe. The region of France, Belgium, and the UK was quite different from how it is today. Sea level was much higher. In addition to the Uralian Sea cutting Europe off from Asia, most of Europe was divided by minor seaways into a chain of archipelago islands. Today, most of Europe and the areas near the Mediterranean are considered temperate environments. However, in the Middle Jurassic, northwestern Europe was a tropical environment. It was much warmer than it usually is now; a hot summer day in modern-day Scotland could reach up to 15 °C (55 °F). In the Middle Jurassic, this would have been considered a harsh winter. The average annual temperature of Europe would have been anywhere between 20–25 °C (68–77 °F), with summers reaching up to 40 °C (104 °F). Rather than large wooded forests, the area was filled with ferns, horsetails, and the earliest cycad and conifer trees. These conifers were similar to the umbrella-shaped and thick-leafed monkey puzzle trees that are commonly found in South America. It is here, in places like the Isle of Skye in Scotland and Oxfordshire in England, that we see the first empire of the dinosaurs emerge.

Sitting at the top of the food chain was *Megalosaurus*, a massive predator reaching up to nearly nine meters (thirty feet) in length. It dwarfed all predatory dinosaurs that came before it. Additionally, it was more muscular and powerful than all the other theropods before it. Many theropods of the Early Jurassic had relatively thin and weak skulls that were optimized for snatching smaller dinosaurs, but the skull of *Megalosaurus* was much larger and thicker than earlier predators, giving it a powerful bite. Its teeth too had become slightly thicker, so as to not break as easily when biting down on its prey. All these features indicated that it had grown beyond small easy catches and was instead focused on hunting the largest prey in its ecosystem. It truly earns its name of "great lizard."

Megalosaurus was not necessarily a unique dinosaur. In fact, it was the namesake of a whole group of predators called the megalosaurs. While they were prevalent in Europe, these predators had a nearly global distribution. Their range spanned all of Laurasia: from modern-day East Asia to the western half of North America. They were even found in Gondwana in modern-day Niger. While they appeared capable of adapting to the different biomes of this new Jurassic world, they preferred the tropics.

Even among such massive predators as the megalosaurs, there were still dinosaurs that towered over them. One such dinosaur was the massive *Cetiosaurus*. *Cetiosaurus*, in particular, was a eusauropod, a more primitive sauropod than the diplodocids. This dinosaur towered above all other dinosaurs in its environment, at sixteen meters (fifty-two feet), and weighed eleven tonnes (twelve tons). It had the iconic long neck and long tail. As one of the tallest dinosaurs on this tropical island chain, *Cetiosaurus* could eat from the canopy without any competition.

Cetiosaurus wasn't the only sauropod to exist in Europe— far from it. Sauropods were the most diverse and successful herbivores of the Jurassic. All environments featured many other breeds of these hulking animals. For Europe in particular, eusauropods were the most common, followed by another group called the macronarians. Groups of sauropods would roam the land near ancient lagoons and shorelines. We can still see their tracks to this day in places like the Isle of Skye. Some of these tracks stretch out uninterrupted for miles. Based on the direction of individual trackways within these groups, it doesn't appear that they lived in herds. Instead, they exhibited what's called cumulative milling behavior. You can see this at a park with birds. Many birds wander around in large patches in search of food, or swarm near someone when they throw food on the ground. These birds may not be closely related to each other, and they don't form a defined social structure. They simply all focus on where the resources are and tolerate each other. The same goes for European sauropods in the Middle Jurassic, like *Cetiosaurus*. Massive groups of these dinosaurs would wander around, likely looking for food.

Among many of these trackways filled with the circular footprints of sauropods, there are also different sets of footprints. These prints are slightly smaller, have three toes, and belong to a two-legged dinosaur. Based on the size and shape, these belong to none other than a megalosaur. These predators commonly stalked the eusauropods of Europe. Even something as big as *Megalosaurus* was still faster than just about any sauropod, with some estimates suggesting megalosaurs could run at thirty-nine to forty-five kilometers per hour (twenty-four to twenty-eight miles per hour). At that speed, they could catch most humans.

It wasn't just these two types of dinosaurs that inhabited Europe. The area was also filled with many unique ornithischians. Large armored dinosaurs with plates on their backs and spikes on their tails walked the coastlines. Fleet-footed neornithischians scampered through the trees. There also existed a group of smaller feathered theropods that hunted what was too fast for the megalosaurs to catch. Not all dinosaurs evolved to be massive or monstrous. These dinosaurs, known as the proceratosaurs, were about the size of dogs and featured large and extravagant fan-like crests. Quick and nimble, these dinosaurs could likely run up to sixty-two kilometers per hour (thirty-eight miles per hour) and had long hands to snatch up small prey. What makes these dinosaurs unique is that they were some of the earliest dinosaurs to be completely covered in early feathers. As we've seen, nearly all dinosaurs had some type of proto-feathers early in their family tree, even if they lost them as they evolved through the Mesozoic. However, we still aren't sure just how much of their body was covered in feathers. The proceratosaurs would have had their whole body covered in feathers. Even though these smaller predators lived throughout Laurasia, which featured an assortment of unique yet warm environments, this landmass still must have gotten cold enough that smaller dinosaurs needed feathers to insulate them. Dinosaurs were clearly adapting to a changing climate, and slowly branched out from their origins in the hot and dry Pangea.

One thing we can learn from Europe during this time is that a side effect of higher global temperatures is sea level rise. It might sound counterintuitive, since heat would evaporate water. However, the increased heat actually melts the ice at the poles of our planet. When this ice melts, it creates more water that enters the ocean, which then raises the sea level. Conversely, when temperatures are much colder, more water begins to freeze and accumulate with the ice at the poles; this reduction in water then lowers sea level.

Sea level during the Mid-to-Late Jurassic Period was significantly higher than it is today due to the warmer climate—so high that much of Europe was underwater and the Uralian Sea cut off Asia from Europe. Vibrant ecosystems filled with corals, clams, and plankton covered many European countries. Fish, sharks, ichthyosaurs, and pliosaurs swam through these ancient marine environments.

As temperatures continue to rapidly increase today, the ancient archipelago of Europe may become a modern reality. According to tide-gauge records, sea level has roughly risen twenty-one to twenty-four centimeters (eight to nine inches) from 1880 to 2020. Nearly half of this rise has occurred in the past twenty-eight years. It doesn't sound like much, but the effects are large when put into a broader context. It's estimated that overall, Europe has lost 970–1,100 kilometers^2 (374–424 miles^2) of coastal land since 1880. However, thanks to dikes and sea walls, this estimate may not be evenly distributed across the entire coast of Europe. Thankfully, most of this land was not urban or developed land; much of it was beaches, wetlands, marshes, and deltas. These areas were already transitional regions close to the ocean that could be easily overtaken. However, these regions are also crucial in reversing climate change for two reasons. The first and most obvious is that the plants that live in these regions directly absorb carbon dioxide from the air. The second reason is that these regions, especially wetlands, bury plant material and restrict how much oxygen this plant material is exposed to, thus slowing decomposition. This slower decomposition helps lock the carbon into the soil. But as sea level rises and overtakes these important yet overlooked carbon sinks, we reduce how much carbon dioxide can be naturally absorbed from the atmosphere.

Land loss is not just a problem Europe is susceptible to. It is a global problem. Wetlands, mangroves, mudflats, and so many more environments around the world are slowly being consumed by the sea. Based on the estimates above, anywhere from 12,600–100,800 kilometers^2 (4,800–39,000 miles^2) of global coastal environments and carbon sinks are now underwater due to rising sea levels. For reference, Sitka, Alaska, is about 12,600 kilometers^2 (4,800 miles^2) in size. Yet it doesn't seem like things are set to slow down any time soon.

Based on current projections set by the European Environment Agency, sea level is set to rise much more in the coming years. By 2100, their estimates range from as little as twenty-eight centimeters (eleven inches), based on decreasing emissions, to as much as 1.02 meters (three feet) if emissions continue to *increase*. That means that 16,800–423,300 kilometers^2 (6,500–163,300 miles^2) could be submerged by 2100. Also, for reference: New York City and the surrounding suburbs are roughly 34,450 kilometers^2 (13,300 miles^2), the entire country of Iceland is about 266,770 kilometers^2 (103,000 miles^2), and Germany is nearly 357,160 kilometers^2 (137,900 miles^2). Entire countries' worth of land could be submerged in less than a century. Vast wetlands that are crucial for counteracting this very rise in sea level will be inundated. Even without taking into consideration the environmental loss caused by rising sea levels, are the dams, dikes, and seawalls that cover our coastal cities capable of handling such a change?

Yet again, nothing is set in stone. Sea level can always come down. Coastal environments can always be rebuilt, especially by caring human hands. The future only seems bleak if we don't use our agency to change our situation, even if only by marginally correcting course.

To the east, across the Uralian Sea, southern China was also a much different place during the Mid-to-Late Jurassic. It was a more moderate landscape by comparison. By today's standards, the coasts were tropical, but inland were warm semi-arid forests. The forests surrounding the riverbeds and floodplains were filled with cycad and ginkgo trees that could handle the more semi-arid conditions. In places like the Yunnan province of China, annual temperatures were similar to Europe; however, it was the most frigid region at the time with winters dropping to 6 °C (42 °F). Hardly a winter wonderland. Furthermore, it was more seasonal in terms of rain: In summer months, it could experience up to eighteen centimeters (seven inches) while the rest of the year, it experienced less than tropical Europe. Korean and Chinese residents who live north of Beijing would have found this ancient climate quite familiar.

Here, a wholly unique ecosystem dominated the landscape. The land was still dominated by the same general kinds of dinosaurs: long-necked sauropods, meat-eating theropods, armored stegosaurs, and swift ornithopods. However, the region boasted a unique set of dinosaurs. The eusauropods were present here, but were not the dominant sauropods. Instead, a group called the mamenchisaurs reigned here. While still long-necked dinosaurs, this group was special in that they had the longest necks of any sauropod ever, with many having necks that made up nearly half of their body length. Some species of *Mamenchisaurus*, in particular, could reach up to thirty meters (one hundred feet) in length and had a neck length roughly twelve meters (forty feet) long.

By far, the mamenchisaurs were the most abundant dinosaurs in the region. It is interesting to note that their environment was quite unique for sauropods. East Asia was uniquely seasonal; the beginning of the year was slightly dry, while the summers were exceptionally high in precipitation. The other regions where sauropods lived were much less seasonal. Additionally, East Asia had warmer spring and autumn temperatures than other regions. This meant that more of the year was warmer, which was quite important for sauropods. Remember: Sauropods were unique in that they were poikilotherms, meaning that, although they could generate body heat, they still relied on the ambient temperature of the environment to provide them with extra heat. Thus, a longer stretch of warmth is exactly what a poikilotherm would want; the longer a region could stay warm, the more likely these types of animals would be able to thrive.

When you look at the long neck of a sauropod, it can be easy to see how it resembles a giraffe. The long neck could easily help it reach up into the tree canopies, where no other dinosaurs could reach. Yet the body of any given sauropod could weigh several tons, far bigger than any land mammal ever. With a body that big, it would require enormous amounts of energy to move; just grazing could be a workout for such an animal. Thus, sauropods likely used their necks to sweep large swaths of ground as well. Instead of moving their bodies to the food, they could stand in one place and move their long necks over the surrounding area, allowing them to take up tons of plant material while expending very little energy. This isn't to say they were lethargic animals; fossil trackways disprove this. They were simply some of the most efficient grazers ever.

Though some species of megalosaurs lived in this region, they also lived alongside the metriacanthosaurs. These predators had small crests above their eyebrows. Rather than muscular arms and jaws, they had skulls and teeth that were optimized for slashing. These specialized skulls would have served them well when hunting the mamenchisaurs, who were large, slow, relatively defenseless, and very meaty. In addition to the larger metriacanthosaurs, the agile proceratosaurs stalked throughout these temperate forests. While these small predators existed in Europe, they had an expansive range in Asia during the Jurassic. These fleet-footed predators could be found all the way from Siberia to China, further demonstrating their ability to live in colder climates.

Across the Pacific Ocean, North America during the Late Jurassic was a whole different environment. What we know of this time mostly comes from the western states: Colorado, Utah, New Mexico, Wyoming, and some of the surrounding states. This region was completely different from how it is today. It was still hotter than today; summers would consistently reach up to 40 °C (104 °F), while winters would hardly reach below 10 °C (50 °F). This area only received less than forty-five centimeters (1.5 feet) of rain a year. Additionally, the Rocky Mountains that cut through most of this region now didn't exist at that point. The tectonic forces that would spring these mountains up wouldn't begin for another seventy million years. All of this together created a vast dry desert that stretched over most of North America, similar to the Serengeti of modern-day Africa. A typical scene in the desert of this region consisted of large riverbeds and floodplains lined with conifer and cycad trees. Even though it was a brutally arid environment, it still could support more fragile life like ferns, horsetails, and amphibians.

In this North American kingdom lived many unique dinosaurs as well. Here, we see many of the iconic species of dinosaurs that we recognize from our childhood. At this point in history, the two-legged sauropodomorphs had gone extinct and their larger four-legged cousins, the sauropods, had become the dominant herbivores of the time. North America was primarily ruled by a group of sauropods known as the diplodocids. You may know a couple of these species: *Diplodocus*, *Apatosaurus*, and *Brontosaurus*. These sauropods had a more vertically oriented posture, rather than a giraffe-like posture, and were known for their unusually long whip-like tails, which were roughly twice as long as their necks. This relatively shorter neck, longer tail, and vertical posture was not very helpful when reaching for tree branches. These sauropods likely focused more on plants that grew close to the ground.

This lifestyle is also evident from their uniquely shaped heads, which are unlike all other sauropods' heads around the world. Most sauropods had short, high skulls as well as spoon-shaped teeth. Dinosaurs like *Apatosaurus* instead had a longer skull with peg-shaped teeth. These peg-shaped teeth were also more concentrated toward the front of their mouth, rather than evenly spaced throughout the mouth; this would be the equivalent of all the teeth in a human skull being crammed into where the canines and incisors are. These peg-like teeth had unique wear patterns that suggest they were efficient at stripping plants. All of this suggests that diplodocids had an entirely different diet than most sauropods. Rather than mostly eating from the conifer and cycad canopies, they would have preferred to graze on ferns and horsetails that grew close to the ground. While diplodocids grazed the plains of western America, this left room for other sauropods to eat from the trees. Macronarians, like the famous *Brachiosaurus*, capitalized on this niche and became some of the tallest dinosaurs in the region and even in the world at the time. These sauropods had massive nostrils situated on the top of their head. It's thought that these massive nostrils helped cool the heads of these huge animals and prevented them from overheating in such warm environments.

Even though the sauropods were a hallmark of this Jurassic desert landscape, many other fantastic animals roamed alongside them. One of them was *Stegosaurus*, the largest armored dinosaur of the time. When we look at the body of *Stegosaurus*, it's hard not to conclude that this dinosaur was well-designed for fighting other dinosaurs. From its neck to its tail ran two rows of plates that could reach up to sixty centimeters (two feet) in height. These plates were in the perfect position to prevent any predator, who were usually taller than it, from biting down from above. Even its throat was covered in tiny pebble-sized bones called ossicles, creating a flexible chainmail-like armor over its most vulnerable area. Moreover, its tail featured the iconic and deadly thagomizer: four spikes reaching up to ninety centimeters (three feet). These spikes made it one of the most dangerous dinosaurs in the Serengeti—enough to kill any predator with a single well-placed swing.

While *Stegosaurus* was about as big as an African elephant, it still paled in comparison to the overwhelming sauropod diversity. As such, it had to live in a way that didn't overlap with the ecology of these dinosaurs; it wouldn't have been able to compete with such hyper-efficient grazers. By looking at its small and dainty skull, we can see that it was a surprisingly picky eater. Usually, generalists have larger and stronger jaws so that they aren't limited in what they can eat and can consume tougher material. This is what many ornithischians were like during the second half of the Jurassic Period. Many ornithischians also had thick teeth that were well-adapted for chewing on strong plant material, much like modern livestock. This thickness was needed because the coarse silica particles in many plants slowly break down teeth. All of this to say that *Stegosaurus* was the complete opposite of what many ornithischians were like. Its skull was exceedingly small: It would have been closer to the size of a rugby ball, which is tiny for a nearly five-tonne (5.5-ton) animal. Its teeth were also small knobs, unlike the thick molar-like teeth better suited for grinding. It's estimated that this behemoth of a dinosaur could only produce a bite force of up to 275 newtons, no stronger than a golden retriever. All this together suggests that *Stegosaurus* had a very specific diet: likely soft plants that grew near the ground, like ferns, horsetails, and mosses. In such a dry environment, these plants would have been found more often near the river's edge or by floodplains.

Although *Stegosaurus* was quite a uniquely-shaped dinosaur, it was not the only armored dinosaur to roam the riverbeds. Many other types of stegosaurs lived here, each with unique arrangements of plates and spikes. In addition to the stegosaurs were another, even less common, armored dinosaur known as the ankylosaurs. These Great Dane-sized animals were essentially like walking tanks; from their heads to their tails, they were covered in bony knobs and spikes, making them even more heavily armored than the stegosaurs. Their generally short and wide body shape kept them low to the ground and protected their soft underbelly from larger predators.

The plains of North America were filled with many other herbivorous dinosaurs that were less striking. The neornithischians were much lower on the food chain than the rest of the herbivorous dinosaurs. They couldn't rely on their size for defense like the sauropods. Neither could they rely on defensive structures like the armored dinosaurs. All they had were their long legs and light frame to keep them from becoming prey. Despite this, they managed to fill a unique niche thanks to their highly specialized teeth. As mentioned earlier, the teeth of neornithischians were more similar to our molars; they were flatter, thicker, and better adapted for chewing harder food. Many herbivorous dinosaurs had limited ways of chewing, or didn't chew at all, which limited what they could realistically eat. Therefore, many neornithischians, like *Camptosaurus*, *Dryosaurus*, and *Enigmacursor*, were able to eat tougher plant material. All dinosaurs were able to find their place within the ecosystem.

The desert environment produced unique predators that seemed better suited to more arid environments. One such dinosaur was *Allosaurus*, the lion of this Serengeti. It was one of the most successful predators of the Late Jurassic. Three species stalked the floodplains, with one being the biggest predator on Earth at the time. Rather than strong muscular arms and powerful jaws, these predators had thinner and more lightly built skulls, like the metriacanthosaurs. This didn't mean they weren't deadly predators. Instead of delivering powerful bites intended for crushing, *Allosaurus* used its jaws and blade-like teeth somewhat like a hand saw; its specialized jaws could open wide at a gape of over ninety degrees, bite down, and pull back to leave massive gashing wounds. This tactic would have been especially effective against the large, fleshy, and relatively defenseless sauropods. Many sauropod bones are etched with tooth marks likely left by *Allosaurus*.

Allosaurus was not the only predator in North America at the time. Just as there was a diverse selection of herbivores of all shapes and sizes, so too was it with the carnivores. Like a unique puzzle piece, each predator played a special role in their environment. One such predator was *Ceratosaurus*, the namesake for the entire group of ceratosaur theropods. Interestingly, the ceratosaurs seemed to prefer slightly hotter, more humid regions of North America than *Allosaurus*. Many cousins of *Ceratosaurus* were common in the southern regions of Gondwana. Not only did these different predators have different climate preferences, the structures of their skulls and arms suggest different bite forces and ways of hunting. The same can be said of the giant megalosaur *Torvosaurus*. These predators were much more common in tropical Europe and had more powerful arms and jaws than allosaurs or ceratosaurs. Thus, these predators may have found ways to avoid competition by sticking to different areas and hunting different prey, something many modern predators do to avoid conflict. Yet, as with anything in nature, nothing is black and white: This region was prone to severe droughts. These predators likely encroached on, fought, hunted, or even cannibalized one another during particularly desperate times.

Not every theropod was as monstrous as these in this region. North America saw an immense diversity of smaller feathered coelurosaurs. Like the wild African dogs compared to lions, these predators were more focused on smaller prey like the neornithischians, sauropod hatchlings, or even mammals. Instead of relying on strength, these predators would have been exceptionally quick. Dinosaurs like *Tanycolagreus* would have been able to reach up to sixty to sixty-two kilometers per hour (thirty-seven to thirty-eight miles per hour), making them some of the fastest dinosaurs on the planet at the time. Some species, like *Ornitholestes*, would have also had long hands that would have been perfect for snatching up smaller prey.

In this new world of shifting continents, the mega-deserts of Pangea were a thing of the past. While much of Laurasia had transitioned out of the harrowing deserts that used to plague Pangea, some places, like North America, still had pockets of deserts. This is because, as the oceans cut through land and expanded, more ocean currents could circulate. This allowed for more land to receive cool and humid ocean air. In general, it's the interiors of continents that are more susceptible to drying out. One of the great things about no longer being a supercontinent is that more land is exposed to the cooling and humid ocean currents. The smaller a continent is, the smaller and less severe these deserts will end up being. No matter how bad global warming gets in our modern era, it will never get as bad as it did in Pangea. It won't be for another 250 million years, when the continents finally reunite, that we'll have to worry about endless inhospitable deserts. However, just because our deserts will be limited in size and only located in the interior of continents, doesn't mean there's no potential for problems. Out of sight does not mean out of mind.

While many nations abut oceans, seas, and other large bodies of water, others are situated in the interiors of our continents. As the planet gets warmer, these deserts are projected to expand. Even with receding coastlines, the extra exposure to currents (albeit less cool at this point) won't be enough to cancel the increasing dryness of deserts. At the very least, deserts will get hotter and drier, forcing people to move to avoid starvation.

Even in the heart of modern deserts, we can find solutions. In fact, companies like Justdiggit have taught the residents of Tanzania how to literally terraform their desert environment. The process is simple: Shallow semi-circular pits called water bunds are dug across an open desert area. The pits are roughly a yard or two apart, and seeded with local desert plants. These pits retain the very precious and rare rainwater and allow it to soak into the ground. You would think that dry land would soak up water like a sponge, but it's quite the opposite. Since the topsoil is dry, it's impermeable, and there is no natural path for water to flow into the ground. Water will just flow downhill and be more likely to flood. Instead, these pits catch this water and force it to soak into the ground. This water then hydrates the resilient seeds in these bunds and provides them with a more hospitable microenvironment. Over just a few years, these large semi-circles grow green with well-hydrated plants. Eventually the plants in the bunds spread and connect to the neighboring bunds. In Kenya and Tanzania, roughly fifty-six kilometers^2 (twenty-one miles^2) of barren desert have been radically transformed into green land—carbon sinks that cool the planet—since 2018. It all sounds like a drop in the bucket, but this is the work of a single company whose only expenses were seeds, shovels, and unskilled labor. How much land can be terraformed when this kind of simple idea scales?

When looking at the Mesozoic, it's quite easy to see all the fearfully great reptiles that had conquered the environment— dinosaurs that were bigger and more alien in appearance than many of the animals that we see today. However, in the quiet shadows of these legendary animals was a group of dinosaurs whose impact far exceeded any other dinosaur. In fact, these dinosaurs have had one of the greatest impacts on science in general.

On the coastal lagoons of the ancient German archipelago lived one species of this group of theropods. It was small, with a long tail, small hooked claws, and tiny teeth. It was unlike nearly all other dinosaurs at the time. It had a tiny pointed snout, with a large head and relatively large eyes. Most peculiar of all was that it was entirely covered in feathers. These weren't like the proto-feathers most coelurosaurs had. From its head down to its tail, it had feathers just like the ones you see on any bird outside your window. Even its dainty clawed arms were covered in large feathers to produce a wing. This was *Archaeopteryx*, one of the first birds to have ever existed.

When fossils of *Archaeopteryx* were first found back in the nineteenth century, they left many early scientists stunned. These fossils were discovered less than twenty years after the name "dinosaur" was even coined. In fact, the first dinosaur to be recognized was *Megalosaurus*. This giant terrifying skeleton suggested that Earth's past was much different from today. Yet, just as people were beginning to realize that the Earth was older than we thought and strange animals lived before humans, *Archaeopteryx* suggested that these massive and ferocious creatures of the past were somehow related to the humble animals of today. Its mixture of dinosaur features and bird features showed that it was a perfect missing link between the two groups. Birds being a feathered miniature type of these ancient reptiles reshaped our entire understanding of evolution and natural history. Even Charles Darwin wrote, "Hardly any recent discovery shows more forcibly than this how little we as yet know of the former inhabitants of the world," when he heard of the *Archaeopteryx* fossils.

Early birds were not just restricted to lagoons in Europe. They lived off the coast of China in swamps and mires as well. Birds were so new and rare during this time period that it's hard to know exactly what kind of climate they preferred. Based on their limited occurrence, birds only seemed to inhabit environments off the coast of Laurasia. Here, they would be far removed from harsh desert plains and likely concentrated in forests. As coelurosaurs, they would have been primed for living in environments that would have been on the colder side for dinosaurs. They had already evolved their warm-blooded metabolism at this point. In fact, *Archaeopteryx* is estimated to have been fully grown in roughly two and a half years. While roughly twice as long as modern birds, *Archaeopteryx* matured faster than most dinosaurs.

Even more impressive than their metabolism was their most apparent feature: the ability to fly. Birds are one of three major groups to evolve wings and fly, the other two being insects and pterosaurs. It is truly an evolutionary marvel. Dinosaurs were already masters at locomotion: Their hips, posture, legs, and air sacs made them highly mobile animals. Yet the ability to fly took this to another level. They had essentially no limits as to where they could travel. They could also take advantage of three dimensions rather than just two; they could move up into the trees to rest, nest, search for food, or even hide from predators.

With flight came a need for a stronger brain; it takes a lot of thinking power to be able to navigate flight. Birds have some of the biggest brains among animals of their same size, and *Archaeopteryx* was no exception. Its brain was significantly more complex than that of any other dinosaur of the time. It had advanced regions that gave it exceptional vision, hearing, and coordination compared to other Jurassic theropods, like *Allosaurus*. Not only do birds have highly developed senses, but they are also quite social creatures. We don't know much about the social behavior of *Archaeopteryx* or other early birds of this time, but considering how similar their brains were to modern birds', a social lifestyle isn't out of the question. Once again, dinosaurs had changed the game. By being concentrated in forests, these unique theropods developed the rare ability to fly.

As we can see, dinosaur diversity exploded in the second half of the Jurassic, even after its significant global boom in the Early Jurassic. Why is it that the Mid-to-Late Jurassic Period saw even more dinosaur diversification? It mostly has to do with the breakup of Pangea. The Early Jurassic saw dinosaurs evolve through means of dispersal, but this time they experienced a different biological process known as vicariance. The splitting continents separated the cosmopolitan population of the dinosaurs. As these continents split apart, populations were cut off from each other by immense oceans and seaways. As we know from dispersal, once populations become isolated, they begin to evolve into different species since they can't interbreed. Thus, the global isolation of dinosaurs spurred unique communities to spring up; the dinosaurs of each continent began to drift apart in similarity. The unique climate of each continent encouraged this by applying distinct pressures not found on the other continents.

Another reason for the exceptional prosperity of the dinosaurs during this time could also be the relatively stable carbon dioxide levels. No major volcanic provinces sprang up during this time, allowing Earth to recover from the eruptions right at the end of the Early Jurassic. Carbon dioxide levels were at their highest at the beginning of the Middle Jurassic and tapered significantly down by the close of the whole Jurassic Period. This allowed forests filled with massive conifer trees and higher seas filled with tiny aquatic plants and algae, known as phytoplankton, the time they needed to convert carbon dioxide into oxygen. Overall, climate change during this time took tens of millions of years, a reminder that climate change doesn't have to be a threat if stretched out over long enough periods of time. The bad luck and harsh times of the Triassic were replaced by unmatched abundance and success during the Jurassic Period. As the Jurassic transitioned into the Early Cretaceous, the unique nature of each continent seemed to be prosperous for some dinosaurs, but disastrous for others.

PART 5

Early Cretaceous: *Reflections in the Past*

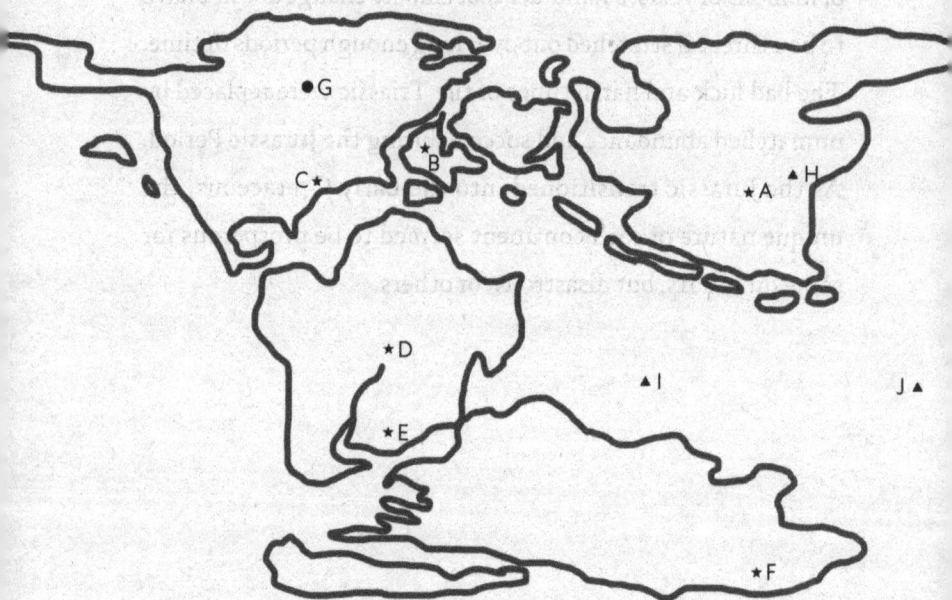

Pangea: ~143.1 million years ago

A - Beijing	C - New York City	E - Johannesburg	G - Carswell impactor
B - London	D - Rio de Janeiro	F - Sydney	H - Liaodong volcanism
			I - Kerguelen Plateau volcanism
			J - Ontong Java Plateau volcanism

Early Cretaceous Dinosaurs

1 - *Yutyrannus*, a proceratosaur
2 - *Suzhousaurus*, a therizinosaur
3 - *Australovenator*, a megaraptor
4 - *Deinonychus*, a dromaeosaur
5 - *Leaellynasaura*, an ornithopod
6 - *Jeholornis*, an early bird

1 - *Cedarosaurus*, a macronarian
2 - *Acrocanthosaurus*, a carcharodontosaur
3 - *Bajadasaurus*, a dicraeosaur
4 - *Iguanodon*, an iguanodont
5 - *Baryonyx*, a spinosaur
6 - *Gastonia*, an ankylosaur

Many paleontologists have noticed how similar the climate of the Early Cretaceous is to our current climate and our most recent past. Overall, it was probably the coldest climate dinosaurs had experienced relative to the hotter climates of the Triassic. Global temperatures would have been roughly 17-18 °C (62-64 °F). The world 143 million years ago wouldn't have felt much warmer than today. Yet it wasn't just the cool climate that had benefited the evolution of the dinosaurs; the long climate trend had also led to this point. From the end of the Triassic to the middle of the Early Cretaceous, a time spanning roughly eighty million years, dinosaurs enjoyed a relatively stable climate. Of course, there were discrete peaks and dips in temperatures throughout this time. However, Earth consistently stayed in the mid-20s °C. Rarely was there a span of time this stable in Earth's history. Even the Jenkyns Event during the end of the Early Jurassic was quite tame compared to the horrific climate disasters that preceded it. Overall, the dominance of the dinosaurs could be attributed in part to this uniquely uniform time in history. As for the beginning of the Early Cretaceous, the global climate continued its trajectory of slowly cooling and staying stable for roughly twenty million years. Earth had struck a unique balance in that it was cold enough in the winters to accumulate ice while still being warm enough to support an abundance of unique plants and dinosaurs. The first half of the Early Cretaceous shows us an ideal for what our modern climate can strive to be like. In the simplest sense, we as humans evolved during a stretch of time that was mostly 15 °C (59 °F) or less, just a little bit colder than the Early Cretaceous. Avoiding a warmer planet is too nebulous a goal without knowing what our baseline is. It also shows just how beneficial a stable climate is, something humanity was a part of before the Industrial Revolution. However, the second half of the Early Cretaceous may remind us how climate change can negatively affect life.

The globe had also begun to look closer to our modern arrangement of continents. During the Early Cretaceous, the northern hemisphere continents continued to shift toward their modern positions; by this point, only Europe and North America were still partially conjoined. After a few million years, Laurasia finally separated, creating the major land masses that we recognize today: North America, Europe, and Asia. However, in the southern hemisphere, Gondwana began to split apart to the extent that Laurasia had back in the Jurassic. It began with South America, Africa, and Antarctica, combined with Australia, India, and Madagascar, opening up to create the very early south Atlantic Ocean. These three regions were only connected by small stretches of land. India, combined with Madagascar, split off from the main southern continent to create an island. Thus, the diversification that took place in the north thanks to vicariance was now taking place among the dinosaurs that inhabited the south. This special arrangement of the continents, combined with the special climatic balance, continued to benefit the evolution of dinosaurs.

One environment that reflects this special climatic balance is China 130–120 million years ago. The ecosystems of this region and time come from the Jehol group of rock formations and are collectively known as the Jehol Biota. The dinosaurs from this specific time and place are integral to the story of Earth's climate history because they reveal how dinosaurs changed from the relatively warmer climate of the Jurassic to a surprisingly colder climate in the Early Cretaceous. This region was filled with vast temperate forests that were colder than those of the Jurassic. On average, this region was roughly 10 °C (50 °F), summers were quite mild, and winters likely dipped into freezing temperatures. The colder temperatures influenced which dinosaurs thrived in China.

The coelurosaurs blossomed here and engulfed the ecosystem. Unlike the metriacanthosaurs that had ruled a few million years prior, now the proceratosaurs dominated this region. Unlike the small and gangly predators of the Jurassic, Early Cretaceous species like *Yutyrannus* reached over seven meters (twenty feet) in length. They were no longer runts, but apex predators. Instead of opting for larger and more extravagant fan-like crests, like in the smaller proceratosaurs, these predators sported a more subtle yet thicker ridge along their snouts. While likely attractive to mates, these skulls were surprisingly well adapted to handle stress, allowing them to deliver stronger bites to prey. But what's most unique about these predators is that—as far as we know—they were the largest fully feathered animals to ever exist. Not with patches, manes, or a vest of feathers, but from head to tail and down to the ankles, these dinosaurs were covered in nearly twenty-centimeter-long (eight-inch-long) feathers.

The skeletons of these massive predators are so exquisitely preserved, you can see the outlines of the feathers in the rock surrounding the whole skeleton. These feathers were more like strands of hair, not quite like the feathers of modern birds. However, they were still effective at trapping heat, which was essential in this frigid environment. *Yutyrannus* shows that even bigger dinosaurs clearly adapted to changing climates. It also reveals that dinosaurs didn't always live in dry deserts, tropical rainforests, or any other type of strictly hot environment; they could live in places we wouldn't normally think of when we think of dinosaurs. Despite having a reptilian origin, these animals could live—thrive, even—in objectively cold environments.

Coelurosaurs weren't just at the top of the food chain; feathered dinosaurs composed most of the fauna in the environment. Their warm-bloodedness kept them from being overly reliant on climate for their success. As evident from the Jurassic, coelurosaurs lived in tropical Europe, moderate East Asia, and the Serengeti of North America; they were capable of living just about anywhere. Now their extremely efficient metabolism was helping them diversify and tap into environments other dinosaurs were less likely to inhabit. Such a trait is what helps modern warm-blooded animals like mammals and birds occupy environments around the world. Coelurosaurs evolved into many different and unique forms to help them cope with the ecosystem they inherited. Dromaeosaurs and troodontids, more commonly known as the raptors, flourished here. With their iconic sickle-shaped toe claw, serrated teeth, and long legs, they made excellent pursuit and ambush predators. However, many evolved unique lifestyles as well. Some smaller dromaeosaurs evolved longer feathers on their legs in addition to their wing feathers, which allowed them to become quad-winged gliders and chase after birds, mammals, and even fish. Troodontids used their extremely sharp eyes and ears to hunt small mammals, but also used their strangely-shaped teeth to eat plants. Besides the raptors, even smaller predators like the one-meter-long (three-foot-long) *Sinosauropteryx* were able to carve out a living here. Usually, small animals don't fare well in cold environments, as they don't have enough body mass to keep themselves warm. However, *Sinosauropteryx* was covered in soft down feathers, nearly identical to those of modern chicks, to efficiently lock in heat.

Not all coelurosaurs were bloodthirsty hunters; in fact, many groups gave up on obligate carnivory (hunting other animals, especially dinosaurs) and embraced herbivory or simply just "less violent" forms of carnivory, like eating bugs or fish. This radical shift in diet spawned many new types of dinosaurs, each with its own unique lifestyle. Some, like *Suzhousaurus*, grew larger hands and claws that were good at hooking tree branches. Others, like *Caudipteryx*, lost their teeth and grew extravagant tail feathers, like modern turkeys, to attract mates. Even some dinosaurs, like *Shenzhousaurus*, developed long legs similar to modern ostriches to help them dart away quickly. And of course, birds dominated the air. This region saw an explosion in bird diversity.

The reign of the non-avian dinosaurs had been impressive up to this point, to say the least. Yet, in the tree tops and in the underbrush, a silent revolution was taking place in birds. Most if not all birds during this time lived in forested areas. Thanks to the breakup of Pangea and the subsequent breakup of Laurasia, the mega-forests that covered these huge land masses had become fragmented as well. Just like the rapid evolution of dinosaurs, birds underwent dispersal, then vicariance, due to the breakup and fragmentation of these forests. Birds began to exert their dominance in the air. During this time, the pterosaurs, who previously dominated the skies, were also going through radical change. Their diversity was expansive in the Jurassic Period. Species like *Pterodactylus* were as common as modern-day sparrows or doves. These smaller pterosaurs hunted for bugs, mammals, lizards, fish, and other small prey. However, in the Early Cretaceous, the vast majority of pterosaurs had evolved to be much bigger; many species of pterosaur were now eagle-sized or larger, and the smaller relatives of *Pterodactylus* went extinct. This abandonment of the smaller niches gave birds the foothold they needed to become smaller flying predators.

On top of that, a new food source burst onto the scene that has been missing for most of our story. The Cretaceous was a special time for dinosaurs, as it saw the origin of angiosperms: flowering and fruiting plants. Flowers and fruit are so ubiquitous today that we take for granted how special this type of plant is. Back then, all other plants required either wind or water to pollinate and reproduce. Seeds didn't fall far from the trees, keeping their range relatively constrained. Only over many generations would plants like conifers, cycads, ferns, and others be able to migrate across regions. With the enticing nectar of flowers and the delicious fruit they bore, flowering plants developed the novel trick of letting animals handle their reproduction via seed dispersal. It worked to amazing effect as flowering plants began to take over the world. Insects rapidly evolved to take advantage of the nectar, spawning many of the pollinators we see today. Animals feasted on the fruits, and the seeds could be transported and sometimes fertilized in the feces of the animal that ate them. This new buffet of options suited birds well. They could now eat small prey, the nectar in flowers, a plethora of new flying insects, or the fruit these new plants bore.

By this point, birds had developed a suite of features to help them adapt to many unique climates. Their air sacs and hollow bones kept their bodies cool in warmer environments and their high metabolism and feathers kept them warm in colder environments. In Early Cretaceous China especially, these feathers had finally become indistinguishable from the feathers of modern birds. Their design allowed for both insulation and lifting capabilities. Flightless dinosaurs leaned into the insulation component. Other groups of ancient birds, like the Enantiornithes, leaned more into the flying capabilities, being born with very few down feathers for warmth. However, the ancestors of modern birds had struck the perfect balance between the two functions of feathers. Now, this combination of new ecological changes opened the door for birds to radically change their physiology to meet these opportunities. The flying capabilities of the Late Jurassic *Archaeopteryx* have always been debated, but this is not the case for birds now. The birds found in the Jehol Biota perfectly exemplify this. Their bones became hollow and extremely light to keep them in the air. Their wings had developed to become more efficiently built. Chest muscles expanded to give them more power in flight. The bony tail was shortened and packed with muscles to improve steering. Even their brains condensed and grew more powerful in order to handle the complex sensory inputs and quick thinking needed to fly. They also developed an array of skull types for different foods. Species like *Confuciusornis* lost their teeth and developed thick triangular beaks to catch fish. Others, like *Longirostravis*, had pencil-thin beaks ideal for probing mud for invertebrates. Some species, like *Hongshanornis*, even swallowed gizzard stones to help them digest plant material. Beyond the Jehol Biota, birds engulfed the world to become some of the most common dinosaurs among ecosystems.

In addition to coelurosaurs and birds, sauropods had changed significantly during the Early Cretaceous. Gone were the mamenchisaurs, which were practically extinct throughout the entire world at this point. Now the macronarians were

the major sauropod group in China. Yet, because of the
colder climate, these sauropods were much smaller; many
species shrank to about ten meters (thirty-three feet) in
length, a far cry from Late Jurassic species who could easily
reach double that length. While still the largest animals in
their environment, this shrank the gap between them and
the ornithischians. The ornithopods specifically began to
grow to much larger sizes. Some species, like *Jinzhousaurus*,
dwarfed their Jurassic ancestors by reaching seven meters
(twenty-three feet) in length. Much like the coelurosaurs,
these dinosaurs were also more warm-blooded in the stricter
sense and could keep themselves warm despite living in a
colder environment. Some smaller ornithopods were known
to have proto-feathers to help keep them warm, but larger
species were still covered in scaly skin.

This will sound controversial, but extinction isn't always
a bad thing. It's not animals going extinct that is a problem;
it's the rate of extinction that needs our attention. Just as
with our climate, change is inevitable. However, how fast our
climate changes is the real concern, because this affects the
evolution and extinction of life in our ecosystems. Climate
change should be slow and gradual, even if it's in a preferred
direction. Extinction, too, needs to happen at a slow and
steady rate. Throughout history, the natural background rate
of extinction is about one species out of every ten thousand
per every one hundred years. This usually isn't a problem, as
roughly the same number of new species evolve at the same

time to replace those that have gone extinct. In some cases, life may even evolve more species than the extinction rate and cause a diversity boom. This was certainly the case for dinosaurs several times already.

Despite it being a wonderful time for dinosaurs in general, dinosaurs did in fact go extinct during the Early Cretaceous. However, for the majority of that period, their extinction rate remained within this natural background rate. This is all thanks to the stable climate of the time, regardless of how hot or cold it was. As we see, many new and better-adapted species evolved to take the place of those that went extinct. This is something we should keep in mind when we talk about extinction in the context of today. For context, there are officially 1.2 million species of animals recognized today. This would mean we would expect about one species to go extinct every year. Anything beyond that and we begin to have problems in the environment. If we don't see any significant climate change in our lifetime, then we shouldn't expect to see any significant extinction events either.

This isn't what we see. The modern rate of extinction in animals is one hundred times the natural background rate. It should have taken eight hundred to ten thousand years to experience the amount of extinction we are seeing in just the past one hundred years. Looking at the past couple years, some species have only been left with a handful of individuals. Without enough individuals to grow the population, some species, like the white rhino, have become functionally extinct. Once those last individuals die, the entire species will be gone. It won't take much more time until we see permanent damage to our ecosystems. Within just three human lifetimes, the amount of extinction we experience may cause some ecosystems to completely fall apart. At this rate, scientists say we are on track to experience a mass extinction known as the Anthropocene extinction. This mass extinction is estimated to be like the one at the end of the Permian, or the one at the end of the Triassic. We know how those turned out. Modern companies, like Colossal, may have found a way to get around this by "resurrecting" extinct animals. Through the use of genetic engineering, scientists can use the preserved genetic material from recently extinct animals, like Dire wolves, and grow them from an embryo into a viable living organism. While this can be used to "de-extinct" animals and rebuild populations, it may not be enough if we don't get to the core issues: habitat loss, deforestation, degrading carbon sinks, increasing carbon emissions, and ultimately global warming. What good is it to bring animals back from extinction if we don't solve why they went extinct in the first place?

It may seem like time is short, but humans are capable of doing unthinkable things in short time spans. Three lifetimes ago was at the start of the nineteenth century. Back then, we didn't have cars, phones, vaccines, nuclear energy, spaceships, or AI. Now we have all of those and more. As we've shown already, it's possible to invest back into carbon sinks and reforest barren deserts in a fraction of that time. It's a heavy lift to reshape a planet's worth of environments back to normal and to reverse our climate change, but if we know our "why," then we can bear any "how."

Much like in China, the land masses that used to be Laurasia became more unique in their climate and underwent slow, gradual changes. In the island chain of Europe, and specifically England, the climate shifted to a more arid ecosystem. Rather than the tropical destination of the Jurassic, it actually began to look similar to today; inland environments transitioned to shrublands like those of the modern Mediterranean. Plant life here began to develop xeromorphic adaptations: changes to their structure that helped them conserve water and energy in a heat- and water-stressed environment. Trees here had thin and small leaves to retain precious water; the larger leaves are, the more surface area they have, which allows more evaporation. The trees here also had highly irregular growth. During drought seasons, they would keep their growth to a minimum to focus their energy simply on survival. The drier conditions prompted dinosaurs to adapt to the changes as well. Across the widening Atlantic Ocean, North America was slowly changing. In the Late Jurassic, most of the interior was filled with dry plains. In the Early Cretaceous, temperatures were steadily cooling, and more importantly, the growing ocean brought more currents to the landmass. As a result, the North American Serengeti shrank. The east coast transitioned to a more humid environment filled with meandering rivers and even swamps. Its connection to Europe was also completely severed at this point.

All of these adaptations not only changed how a plant survived, but also affected how palatable certain plants were. Plants used to more humid climates are softer and easier to break down, while plants in drier climates become tougher. This caused other plant eaters who were capable of eating denser leaves to diversify. A group of ornithopod dinosaurs known as the iguanodonts became exceedingly prevalent in the northern hemisphere during this time. These dinosaurs, named after the iconic *Iguanodon*, mostly had a beaked snout and a unique set of front feet that were hoof-like and bore a thick thumb spike for defense.

Iguanodon was unlike the sauropods that ruled the Jurassic Period. For starters, this European dinosaur was only about nine meters (thirty feet) in length. While big for an animal, it couldn't rely on unlimited options for food like truly enormous sauropods. However, it succeeded in its environment thanks to its unique skull. Rather than spoon-shaped teeth good for stripping plants and digesting them extensively in the gut, *Iguanodon* had short blunt teeth, similar to our molars, and a thick and powerful jaw, all thanks to its neornithischian ancestry. This made it capable of chewing and breaking down some of the tougher plants that made their home in this shrubland. Herbivores that chewed their food were not that common in the Jurassic; even in North America, the fossil record is dominated by the relatively dainty skulls of dinosaurs like *Diplodocus* and *Stegosaurus*. Ornithopods were not the hallmark herbivore in most environments, despite their efficient jaws. Now the iguanodonts began to rise up as the dominant mid-sized herbivores. While they were most notable in Europe, iguanodonts inhabited the entire globe, into North America and Asia as well, their sturdy jaws coming in handy across many types of environments.

Ornithopods and iguanodonts weren't the only ornithischians to flourish. The ankylosaurs, which had originated in the Jurassic alongside their more prominent stegosaur cousins, exploded onto the scene. Though armored dinosaurs, they were quite different from dinosaurs like *Stegosaurus*. One aspect was that they were significantly more armored; rather than only having two rows of plates on their back, these dinosaurs were covered from head to tail in bony armor. Their tails featured an assortment of weaponry. Rather than the thagomizer of stegosaurs, ankylosaurs boasted beefy clubs, stiff maces, or axes. The only area left unprotected was their soft underbelly. Thanks to their relatively flat and low-lying body, the underbelly was kept close to the ground, making it practically impossible for any predator to actually wound one of these dinosaurs. Surprisingly, it was this unique body style that made them excellent low browsers.

Despite having heads optimized for maximum protection, they also were exceedingly complex underneath all the armor. Their teeth and jaws were simple, like stegosaurs. However, these dinosaurs also had long, flexible tongues, which may have assisted them when chewing, or, like giraffes, may have grabbed food. Moreover, thanks to the reinforced structure of their skulls, they could handle greater stress and better chew their food. Many ankylosaurs also had unusually long nasal passages that swirled inside their skull. These long, convoluted nostrils helped air circulate within their head and keep their head cool. It also was effective in capturing moisture in the air. While their warm-blooded nature as ornithischians allowed them to inhabit colder environments, this unique air conditioning and moisture retention system helped them inhabit warmer and arid climates too. This is exactly what happened as ankylosaurs began to spread across all the major continents from the North Pole to the South Pole. Stegosaurs, on the other hand, began to slowly decline. It's doubtful that the delicate skulls of these armored dinosaurs had the same adaptations. Combined with the cooling climate and changing flora, these dinosaurs became a relic of the Jurassic.

The shifting dynamics of the ecosystem were also observed in how sauropods changed. While still a staple in the environment, they became much less prevalent. Sauropods were no longer as diverse as they were in the Jurassic Period. Indeed, sauropods that primarily fed on low-lying and medium-sized flora, like the diplodocids, became increasingly rare. Other older types, like eusauropods and mamenchisaurs, slowly went extinct, leaving only the canopy-browsing macronarians as the main groups of sauropods. Another group of sauropods began to originate during this time as well. These sauropods, called the titanosaurs, were truly unique among the long-necked dinosaurs. They had even stronger limbs to support bigger bodies. However, as predators grew increasingly bigger, size became less of a deterrent for sauropods. These sauropods evolved new ways to defend themselves straight from the playbook of the ankylosaurs: armor. Many titanosaurs commonly grew large pairs of spikes running from their hips to their tails to offer some protection from a pursuing predator. Some species took it a step further and developed tiny pebble-sized bones speckled across their back. Like the chainmail of knights, this provided a flexible assortment of tough bones that didn't bog down the titanosaurs. Yet, despite being a more evolved sauropod, they were still not extraordinarily diverse during this time period.

While the decline of a global group of animals, like sauropods, and the rise of another, like ornithischians, is complicated to say the least, there may be a few explanations as to why. One is most definitely the relatively simple jaws and teeth of sauropods. In a Jurassic environment, simply stripping leaves and processing them in the gut, rather than chewing them, worked just fine. This also allowed them to eat the plants that were relatively easy to digest in bulk. Ornithischians, on the other hand, were developing more complex jaws and teeth that opened them up to more options. The second reason may be the increased variation in environments. The freshly-split continents did indeed have unique environments, which is why vicariance became so prevalent in the Jurassic. However, environments were still overwhelmingly dominated by conifers and cycads, which were exactly what sauropods had evolved to eat in bulk. Yet, as the continents drifted farther apart, local climates differentiated, and environments continued to diversify, the plant life diversified as well. Environments were not entirely dominated by conifer and cycad trees, leaving relatively less food for the sauropods to eat. Additionally, some environments, like England during the Early Cretaceous, became more arid. This made plants tougher to chew and thus less appealing to sauropods and more appealing to ornithischians. The result was that the dynasty of sauropods was beginning to fade, and a new empire of herbivores was rising up.

Climate change affects animals at a fundamental level by affecting the plants they eat. Plants are the foundation of the food web, and their lifestyle trickles down and affects the animals that feed on them, which trickles down and affects the predators that eat those herbivores. As today's environments become increasingly warmer and more arid, plants will respond by becoming more xeromorphic. Leaves will become thinner and more compacted to retain moisture. The sizes of plants will also decrease in order to save resources. While we have a multitude of prehistoric examples to show this, modern scientists are able to observe this right now.

Near Mount Carmel in Israel, scientists have been studying a unique canyon environment that features several unique microclimates based on its shape. This "Evolution Canyon" has shown since 1991 how plants have adapted to different climate stressors. The southern slope of the canyon has hotter and more arid conditions, and the northern slope of the same canyon has relatively cooler and more humid conditions. Even though the plants on each slope are within walking distance of each other, they have evolved to look significantly different. In fact, the plants of the hotter and drier slope of the canyon look more like plants found in the African savannah, while the plants on the cooler and humid slope look more like plants found in Europe. The heat-stressed plants indeed had smaller and denser leaves, and their shape changed as well. Even the genome of these heat-stressed plants diversified, likely due to genetic mutations that helped them survive in the harsher conditions. The changes didn't stop there; while the team mostly studied how plants adapted to climatic changes, they also observed that the flies on the warmer slope made a massive migration over to the cooler side to survive. Scientists studying the Cape Floristic Region of South Africa were able to see a very similar result. This steep mountain range also had distinct microclimates on the western and eastern slopes. Again, on the slopes that were hotter and drier, plants had evolved for heat-stressed conditions by developing smaller and denser leaves. These microclimates essentially recreated the forces of natural selection that Darwin realized in the Galapagos when he formulated his theory of evolution; changes in the environments have cascading effects on the life that inhabits them, and these effects create distinct features and species.

While we can see the effect of climate change right now on plants, these microclimates are few and far between. One could argue that these may simply be evolutionary novelties found in remote and rare regions. How do we know this actually predicts what will happen in a future affected by global warming? Scientists in Germany sought to answer that exact question. By growing a broad selection of trees in a controlled environment, they could raise the temperature high and see how well adapted these trees were. Some trees could barely handle 35.4 °C (95 °F). Others could manage to survive all the way up to 53.6 °C (128 °F). The big difference between the trees that could survive the extreme temperatures and the species that couldn't was that they already had leaves that were thicker and thinner than average, respectively. These were all trees that live in the temperate environments that humans do. Thus, as temperatures increase over the years, we will begin to see a change in composition. Our forests will slowly lose broad-leaved trees, like ash trees. As a result, wildlife will have to adapt to the changing flora composition. Animals with stronger jaws and teeth that can grind and chew tougher plant material will be more likely to handle such a dramatic environmental change. However, animals that have evolved to eat softer plant material will either be forced to migrate or go extinct.

Climate change doesn't exclusively affect plant-eating animals, however. Carnivores must contend not only with the changing climate, but also with significant change in the prey they hunt. Many of the prominent groups of theropods that ruled over Jurassic ecosystems went through radical changes during the first part of the Early Cretaceous as well. The megalosaurs who dominated the tropics of Europe and northern Africa had all but faded away except for a group of hyper-specialized cousins. These relatives, known as the spinosaurs, had many features that the megalosaurs had: large claws and skulls that made them fearsome predators. However, the spinosaurs were caricatures of their cousins. Their skulls became elongated to the point where they looked surprisingly similar to crocodiles. Their large claws became even more exaggerated as they grew in size, and more hooked. The names of certain spinosaurs, like *Suchomimus* ("crocodile mimic") and *Baryonyx* ("heavy claw"), point to this exaggeration of features. Yet the unique body of the spinosaurs fit perfectly into this new landscape. Although they lived in much more arid environments than their ancestors, they preferred coastal and river environments. These predators used their long snouts and hooked claws not for hunting other dinosaurs, but primarily to snatch up fish along shorelines and rivers. Some may have even used their dense bones to dive deep beneath the water to catch fish.

Megalosaurs were not the only theropods to go through a significant change in the Early Cretaceous. The allosaurs went extinct too, but a relative of theirs rose to dominance, known as the carcharodontosaurs. The name of these predators translates to "shark-toothed lizards," which alludes to their thin, sharp, and serrated teeth that were adapted for tearing meat. Like their Jurassic relatives, this made them predisposed to hunting down the large, fleshy sauropods. As sauropods got bigger, these carcharodontosaurs got bigger as well. Some species, like the North American *Acrocanthosaurus*, could reach eleven meters (thirty-six feet) and weigh up to 3.6 tonnes (3.9 tons), making them some of the largest predators to ever exist.

With spinosaurs sticking mostly to lakes and rivers and carcharodontosaurs focusing on the big game sauropods, this left many open niches for the coelurosaurs to fill. Across the globe, feathered theropods became an integral part of ecosystems. Much like in China, their diverse diets and ecology allowed many different species the opportunity to flourish. The growing success of the coelurosaurs across many Early Cretaceous environments likely had to do with their bird-like metabolism. Like birds, they were both endothermic and homeothermic, meaning they could warm themselves up *and* maintain a constant body temperature. This was especially helpful in places like East Asia, where they could keep themselves warm despite the freezing temperatures. It also helped them cool themselves down and allowed them to inhabit hotter and drier environments, like prehistoric England and the American Southwest, if the opportunity arose. The same couldn't be said for the old guard of theropods; spinosaurs and carcharodontosaurs began to increasingly prefer hot and dry climates. This preference left fewer environments for these giant theropods to inhabit.

Additionally, the more warm-blooded nature of coelurosaurs meant that they grew faster, and thus reached their mature size sooner. Many species reached adulthood around age ten, whereas other older breeds of theropods could take several decades to do so. Carcharodontosaurs were in fact the longest-living theropods. This was an evolutionary game-changer. This meant that coelurosaur predators could begin hunting prey sooner, and even hunt bigger prey sooner. They also spent less time being a vulnerable target as a juvenile. Finally, they could reach sexual maturity and begin reproducing faster. Acquiring resources, not dying, and passing on your genes are the most important ways to secure evolutionary success. Thus, coelurosaurs achieved a level of success other theropods could never attain.

The long stability of the Mesozoic would eventually end. The dinosaurs would face some of the worst climate change they'd ever experienced in millions of years. Beneath the shifting continents, magma began to shift, rise to the surface, and create volcanic provinces around 121 million years ago in a span of time known as the Aptian Age. It wasn't just one plume. Three separate provinces grew. The first was located in the Liaodong Peninsula in China. The second was located in the Kerguelen Plateau, which sits triangulated between Antarctica, Australia, and South Africa. The final and largest was the Ontong Java Plateau, which was northeast of Australia in what is today the Solomon Islands. The Ontong Java Plateau was so big, in fact, that it's still around today; the outpouring of lava from this province created a vast expanse of seamounts and the largest atoll on Earth. This atoll is even home to several thousand Polynesian people today. While not connected and in different regions of Earth, these three volcanic provinces all erupted at the same time. After it was all said and done, roughly 3.5 million kilometers2 (1.3 million miles2) of lava erupted between them during the Aptian. This made the impact of this event nearly as large as the CAMP eruptions, which saw the destruction of nearly all non-dinosaur animals at the end of the Triassic. It's estimated that global temperatures skyrocketed about 4 °C (39 °F) in roughly a million years, a radical change given how quickly this happened.

During this global warming, we get a glimpse of how life in South America and Africa adapted to the change. Most of the landmass became a harsh desert, thanks to its closeness to the equator. The northern regions, like Morocco and Brazil, were tropical, thanks to the widening Atlantic Ocean. However, like North America in the Late Jurassic, the vast interior, cut off from ocean currents, produced an arid landscape. Most regions were over 30 °C (86 °F) on average throughout the year. Precipitation nose-dived, making it one of the hottest and driest places on Earth.

Rather than become nonviable, new and unique breeds of dinosaurs were able to handle the intense aridity and thrive. In fact, many descendants of dinosaurs that lived in the Late Jurassic of North America thrived here. While the vast majority of diplodocids went extinct, a small sect of them proliferated in the southern hemisphere during this brief period. Known as the rebacchisaurs and the dicraeosaurs, these descendants of the drought-adapted sauropods spread across the landmass. Like many direct descendants, they accentuated the features that made diplodocids so successful. The skulls of these herbivores became boxier and their teeth became long and thin; this gave their jaws the appearance of a comb, optimized for scraping up the soft and nutritious ground cover. At the same time, many of these dinosaurs seemed to forsake the very feature that all sauropods have: the iconic long neck. Many of these sauropods had shorter necks, further suggesting they preferred low-lying plants rather than tree canopies. Other strange adaptations came in the form of massive neck fans among certain dicraeosaurs. Species like *Bajadasaurus* sported nearly fifty-eight-centimeter-long (two-foot-long) spines along their necks that acted as struts for large, fleshy sails.

In such a hot and dry climate, how did massive animals like these sauropods survive? They relied upon the same trick many modern animals use to survive the same conditions. African elephants and giraffes are some of the biggest land animals on Earth, yet they live in the hot African savannah. Rather than overheating, they can cool themselves down extremely well thanks to their body shapes. Animals can warm themselves up more easily the bigger they are, because they have more volume. This greater volume traps heat. On the flipside, heat dissipates through the skin, which is on the surface of the animal. Therefore, the more surface area an animal has, the better it can cool itself down. The long neck of the giraffe and the large ears and trunk of the African elephant help increase their surface area and keep them cool in such a hot environment. Sauropods used the same trick: Their long necks and tails provided ample surface area that could keep their massive bodies cool regardless of where they lived. As for the short-necked dicraeosaurs, instead of relying on long necks, they likely got around this thanks to the massive sails on their necks.

During this episode of global warming, we see the ocean become increasingly anoxic, meaning less oxygen is held in the ocean. As a liquid becomes warmer, it has a hard time holding onto gases; the increased temperature makes the molecules move more, which mean increased opportunities for gases to wiggle out of the liquid and into the air. For example, soda has carbon dioxide dissolved into it, which is why it's fizzy. A fresh ice-cold soda is very fizzy when opened, but a room-temperature or even warm soda is not as fizzy when opened. Furthermore, if you keep an opened soda in the fridge, it will stay fizzy and fresh longer, but a room-temperature soda gets flat quickly.

Dissolved oxygen levels in the ocean are essential for the health of the ocean; even though fish live in the water, it is this dissolved oxygen that they extract with their gills to breathe. Thus, as oceans warm, fish and other sea creatures slowly suffocate. That's not all. Remember: As global temperatures increase, the intensity of the water cycle increases with it. With more intense storms and more flooding, this means that the land experiences more erosion. As erosion increases, more nutrients from the continents make their way into the ocean, where they are consumed by phytoplankton, creating massive algal blooms: huge green swaths on the surface of the ocean. Wouldn't this be good, though? Phytoplankton are a major carbon sink that converts carbon dioxide into oxygen. That's good for cooling the planet, right? It is momentarily, but the consequences of algal blooms exceed their benefits. Plankton have a very short lifespan, and these phytoplankton die and sink to the bottom of the ocean, where they decompose. The microbes involved in decomposing these plankton break down the organic matter while using oxygen to create energy for themselves and carbon dioxide and water as waste byproducts. Thus, oxygen is reduced and carbon dioxide is increased every time plankton is consumed. Under normal conditions, this isn't an issue, but the sheer volume of algal blooms ends up depleting ocean oxygen while also creating an abundance of carbon dioxide. To make matters worse, the magma from the Liaodong Peninsula, Kerguelen Plateau, and Ontong Java Plateau volcanic eruptions cooled into volcanic rock, which was then eroded and introduced additional nutrients into the ocean. This then provided even more food for algal blooms to grow, exacerbating the problem. As a result, oceans around the globe became inhospitable on a level unlike anything thus far in Earth's history. Sea life began to suffocate across these new oceans. While sea life was hit hardest, the dinosaurs on land were not spared from the effects. The carbon dioxide buildup from this widespread ocean anoxia added to the carbon dioxide being released from these eruptions to create an increasingly hothouse environment.

The oceans becoming anoxic wasn't a one-time incident during the Aptian; anoxic events are common throughout prehistory, and are even more common during episodes of global warming. In fact, anoxic oceans are still a concern today, based on global warming trends. The Gulf of Mexico is one body of water that is significantly anoxic. Over 7,770 kilometers2 (three thousand miles2) of the gulf is considered a dead zone, where marine life cannot survive. Another anoxic region is the Baltic Sea in northern Europe. Both of these regions have become anoxic from the nutrient-rich runoff from agriculture that feeds algal blooms. What's more, their geography as semi-enclosed bodies of water means that mixing in more oxygenated water is less likely. As temperatures increase, oxygen will escape into the atmosphere and greater erosion will likely increase nutrient-rich runoff, growing these anoxic regions. There are also bodies of water that are on the verge of becoming anoxic. Regions like the Tropical Eastern Pacific (located off the west coasts of California, Mexico, and Central America) and the northern Indian Ocean (located near the Arabian Sea and Bay of Bengal) have oxygen levels that are dangerously low for life. As temperatures increase, we could see these regions become anoxic. Luckily for us, the anoxia in our modern oceans is nowhere near the anoxic events of the Early Cretaceous; the fact that they are contained to certain regions of the ocean is a relief—at least for now. However, it is a warning sign for what may be in our future. If global warming continues, it's likely that these dead zones will expand or new ones open up.

Just as suddenly as the global temperature grew, Earth went through an abrupt cold snap only a few million years later, still during the Aptian. The culprit for this global cooling likely doesn't involve any volcanic ash from these provinces. Instead, this cold snap coincides with the impact of yet another meteor, known as the Carswell impactor. This meteor touched down in northern Saskatchewan, Canada, and left a crater thirty-nine kilometers (twenty-four miles) in diameter. The crater was less than half the size of the Manicouagan crater from the Late Triassic; therefore, the impact winter would have been less severe. However, it was still enough to quickly reverse the trend of global warming.

Dinosaurs went from blistering summers right back into freezing winters. These temperatures weren't that different from what they were before the most recent global warming. However, the rate at which the planet cooled down was a big problem. Ice began to accumulate at the poles, forcing dinosaurs to rapidly reacclimate to frigid forests. Strangely enough, during this cold snap, Australia became home to many unique polar dinosaurs. Australia at this time was a far cry from what it is today. Instead of expansive scorching deserts, this continent was situated at the South Pole, and at times was still connected to Antarctica. During the Aptian cold snap, this region was around 2-4 °C (35.6-39.2 °F) on average, but winters frequently reached far below freezing. Australia's position near the South Pole meant that dinosaurs here would go through winters in which the sun wouldn't rise for weeks. The darkness and freezing temperatures bred a unique group of dinosaurs. The most common herbivores here were small ornithopods, only about the size of turkeys. Small species like *Leaellynasaura* scampered through the conifer and cycad forests, eating soft low-lying plants. Even though temperatures reached freezing for much of the winter, the summers were still warm and wet enough to allow the forest floors to be filled with ferns and horsetails.

In a landscape where winters were harsh and herbivores were roughly twenty-two kilograms (ten pounds), the predators of this region had to adapt in unique ways. Enter the megaraptors. Unlike their misleading name suggests, they were not relatives of raptors. These six-meter-long (twenty-foot-long) carnivores are thought to be coelurosaurs and relatives of the proceratosaurs that lived in Asia. Instead of having massive curved claws on their toes like raptors, these predators had large hooked claws on their hands. Rather than having powerful reinforced skulls to deliver deadly bites, predators like *Australovenator* had thin slender skulls that were better for slashing bites. This unique combination wouldn't have been especially effective against big game, but was perfect for the quick, tiny ornithopods of Australia. Megaraptors would have used these large powerful arms to easily hook and pull in these ornithopods. With its catch secure, there was no need for crushing bites, as it ate holding its prey in its hands. As a coelurosaur, its warm-blooded nature and massive size made it better adapted to colder climates. While megaraptors were scattered around the globe, they were concentrated in the ancient South Pole in places like the southern tip of Argentina and Australia. No other large predator of Gondwana seemed to handle living in such cold environments, leaving megaraptors the undisputed kings of the South Pole.

During harsh winters, many modern mammals hibernate to save energy and stay safe during the cold temperatures. However, reptiles don't hibernate; in fact, most don't even live in places that get this cold. So, how did dinosaurs survive in such freezing temperatures? The answer can be found in the bones of these polar dinosaurs. Remember: By looking at the spacing of the growth rings within dinosaur bones, you can understand how quickly or slowly they grew. You can also tell, based on changes in the composition of the bone, if they underwent seasonal changes in growth. Australian polar dinosaurs didn't seem to slow their growth as if they were hibernating. In fact, they grew just as quickly as dinosaurs at lower latitudes. There also seems to be no evidence that they seasonally migrated north for warmer weather. It seems that dinosaurs simply stuck it out during the winter and survived the freezing temperatures.

The sudden warming of the Aptian and the cold snap of the Aptian-Albian was not the last of the climate change. Once again, temperatures began to rise, and ocean anoxia began to intensify as a result of the continued eruptions of these large igneous provinces. The eruptions from these provinces were so intense and generated so much new crust that new tectonic plates were created in the Pacific Ocean. These new tectonic plates even began to separate India from the rest of Gondwana and send it drifting toward its future home in Asia. Even after this sudden cold snap, temperatures quickly returned to the balmy climate of just before.

All of Earth was affected by this global warming. Around the world, ecosystems that each had regional differences uniformly got hotter and more humid. It appeared as if Earth began to shift to a globally tropical climate. For some environments, this tropical shift was not as abrupt, but for some it was a radical change. No place on Earth was hit quite as hard as Australia. In the span of ten million years, average temperatures skyrocketed 15 °C hotter, the greatest change in temperature among any region on Earth at the time. Additionally, it experienced significantly more precipitation. By the end of the Early Cretaceous, it experienced nearly three times as much rain as before. What was once an icy forest had dramatically changed into a completely different environment. This is evident by the change in dinosaurs that inhabited Australia. Sauropods, which were very rare in Australia in the Early Cretaceous, especially in the cold snap, slowly became more and more common. Global warming raised the temperatures enough to allow many sauropods in South America to cross through Antarctica and into Australia.

This dramatic change may seem like a strange phenomenon of the prehistoric world, but in fact it's something we can observe here in our modern time. Antarctica is currently heating up roughly four times faster than average. Why? Antarctica currently is mostly covered in thick ice sheets. The white ice and snow actually reflect heat. But as temperatures increase, more ice melts, and thus less heat is reflected back. This creates a positive feedback loop. That's not all: Ocean water is much better than ice at absorbing heat. Therefore, as more ice melts into water, more heat can be absorbed, which further increases temperatures. This creates another positive feedback loop. Less ice means less heat reflected and more water. More water means more heat absorbed and less ice. This Arctic amplification means that the poles are significantly more vulnerable to climate change than many other regions on Earth.

The growing sauropod diversity of Australia is an interesting example of one thing dinosaurs were able to do to avoid or take advantage of climate change: move. Although the continents had already begun to split apart, with much of Laurasia separated into our modern landmasses, the relatively short distances between continents meant that land bridges could frequently open up. Dinosaurs in Gondwana were able to easily migrate across what would be Africa, South America, Antarctica, and Australia if the climate prompted them to. There even existed ancient land bridges between Europe and Africa. Near the very end of the Early Cretaceous, we begin to see hints of African dinosaurs in Europe, likely chasing their climate preferences as the equator heated up. There were also land bridges between Asia and North America; the famous Bering Strait that facilitated the migration of humans was indeed available to dinosaurs as well. While dinosaurs faced a rapidly changing climate, species had the option to move to a more preferable climate rather than adapt to the changes.

Animals nowadays don't quite have that luxury. As the continents continued to drift apart for the next one hundred million years, these land bridges began to disappear. Today, only two major land bridges still exist: the Isthmus of Panama bridges North and South America, and the Sinai Peninsula bridges Africa to Asia. There are no land bridges for Antarctica and Australia, leaving the inhabitants forced to adapt to the changing climate. The Bering Strait no longer exists, leaving the hemispheres isolated from each other. The many islands of the Pacific Ocean are isolated as well. This is the unfortunate downside to the separation of the continents; while new and unique climates spring up to foster evolution, life is as geographically separated as it has ever been. Few ecosystems will be able to migrate. If sea levels rise, these few land bridges may become submerged as well, keeping some regions entirely isolated. Some species may not be able to adapt to the changing climate and go completely extinct.

Extinction is a fate many dinosaurs began to face as temperatures climbed at the end of the Early Cretaceous.

Stegosaurs, once titans of the Jurassic, had fizzled into extinction by the end of the Early Cretaceous. Their delicate skulls and particular diet likely did not mesh well with the increasingly arid climate shift. However, many other dinosaurs were now threatened by this climate change. As temperatures continued to soar, dinosaurs were thrown into a climate they hadn't seen in millions of years. Indeed, the hothouse climate of the Late Cretaceous was very similar to the magnitude of heat that dinosaurs faced in the Late Triassic. Now, with the shifting continents, new ocean currents, and unique regional differences, things would play out much differently than they did millions of years ago. They were on the brink of a radical new change.

The climatic changes would not only have been familiar to dinosaurs, but to humans as well. During the Aptian-Albian cold snap, the average global temperature was roughly 19 °C (66.2 °F). This isn't much higher than today's global average temperatures. However, in the Albian, temperatures began to slowly increase by roughly 2 °C every million years. Many paleontologists have noticed how eerily similar this temperature increase at the end of the Early Cretaceous is to our current global warming situation. Thus, the events of the Late Cretaceous may be a glimpse into what the future of humanity might look like as global warming affects our planet.

PART 6

Late Cretaceous: *Pax Dinosauria*

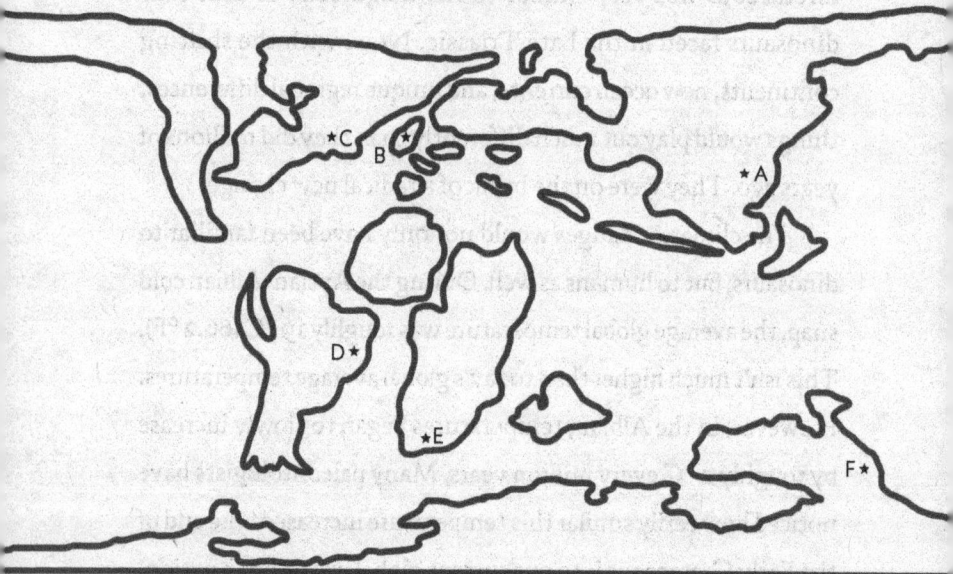

Pangea: ~100.5 million years ago

A - Beijing   C - New York City   E - Johannesburg
B - London   D - Rio de Janeiro   F - Sydney

A s dinosaurs transitioned into the second half of the Cretaceous, their climate began to reach a boiling point. Due to the recent eruptions of the large igneous provinces

Mid-Cretaceous Thermal Maximum Dinosaurs

1 - *Argentinosaurus*, a titanosaur and one of the largest land animals to exist
2 - *Spinosaurus*, the largest spinosaur
3 - *Giganotosaurus*, the largest carcharodontosaur
4 - *Eolambia*, an early hadrosaur
5 - *Moros*, an early tyrannosaur
6 - *Zuniceratops*, an early ceratopsian

in the Early Cretaceous, carbon dioxide levels were about five times greater than they are today. While estimates vary based on the proxy method, they could be anywhere from two thousand to five thousand parts per million—some of the highest in Earth's history. Global temperatures kept rising after the Aptian cold snap for roughly ten million years, at a terrifying rate. Around ninety-three million years ago, in the middle of the Cretaceous, they hit their peak at 28 °C (82.4 °F), throwing dinosaurs right into a hothouse climate dubbed the mid-Cretaceous thermal maximum. Not since the cataclysm at the end of the Permian had Earth been this hot, and in fact we never see Earth return to such extreme

Late Cretaceous Southern Hemisphere Dinosaurs

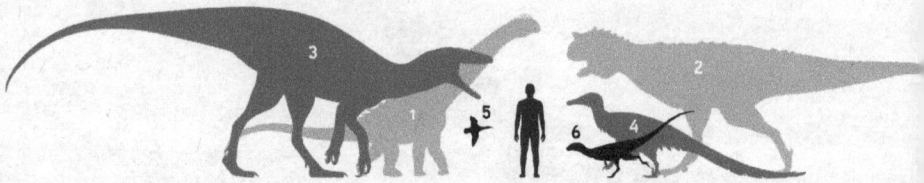

1 - *Titanomachya*, a titanosaur
2 - *Carnotaurus*, an abelisaur
3 - *Maip*, the largest megaraptor
4 - *Austroraptor*, an unenlagiine
5 - *Vegavis*, an early modern bird and ancestor to waterfowl
6 - *Masiakasaurus*, a noasaur

many dinosaurs to go through a dramatic crescendo in diversity. In places like Morocco and Egypt, more ancient types of dinosaurs, like spinosaurs, carcharodontosaurs, rebbachisaurs, and titanosaurs, became more numerous and increased in size. Predators in this region became some of the biggest carnivorous animals to ever walk the Earth. Within the inland arid deserts, hunters like *Tameryraptor* and *Carcharodontosaurus* could reach up to twelve meters (thirty-nine feet) in length. Yet across the mangrove coastlines, *Spinosaurus* dwarfed them all at nearly fifteen meters (fifty feet) in length. Never had a theropod been able to reach such staggering sizes. Even sauropods, like *Paralititan*, reached stunning lengths of nearly twenty-seven meters (eighty-nine feet). Despite these types of dinosaurs being on the decline since their ancestral reign

Late Cretaceous Northern Hemisphere Dinosaurs

1 - *Edmontosaurus*, a hadrosaur
2 - *Tyrannosaurus rex*, the largest tyrannosaur
3 - *Triceratops*, a ceratopsian
4 - *Ankylosaurus*, the largest ankylosaur
5 - *Asteriornis*, an early modern bird and ancestor to gamefowl
6 - *Anzu*, an oviraptor

in the Jurassic, non-coelurosaur theropods and sauropods took advantage of the change.

While the success of these older types of dinosaurs was evident across the globe, Argentina was the epicenter of their rule. Carcharodontosaurs, like *Mapusaurus*, still dominated as terrifying predators, but they were put to shame by the titanosaurs of the region. Towering above all other life were dinosaurs like *Argentinosaurus*. This dinosaur in particular reached up to thirty-five meters (115 feet) in length and weighed a gargantuan ninety-four tonnes (104 tons). This single animal was greater in mass than a herd of African elephants. It was the largest dinosaur—the largest animal—to ever walk the Earth in all of history. Not only did these dinosaurs exact their dominance in size, but they were also more numerous than ever. Patagonia ninety million years ago saw at least four unique species of rebbachisaurs, four species of titanosaurs, and three species of carcharodontosaurs; the sheer diversity of such gigantic animals, in addition to the wealth of other more "normal"-sized dinosaurs, was truly unique.

The global warming of the Middle Cretaceous selected for certain dinosaurs that already thrived in warm environments. Remember that in the Early Triassic, when temperatures were similar to this, it wasn't the more warm-blooded dinosaurs that were on top. In the immediate aftermath of the end-Permian extinction, it was the mesothermic or even ectothermic archosaurs that dominated Pangea. This biological principle held true in the thermal maximum as well. While not ectotherms anymore, dinosaurs with relatively lower metabolisms would be more successful than dinosaurs with higher metabolisms. This is why spinosaurs, carcharodontosaurs, titanosaurs, and rebbachisaurs began to prosper during this climate crisis. True, a lower metabolism meant that in a colder climate you would grow slower, not get as big, and not be as active because these demanded more energy. However, when a warmer climate comes along and can artificially raise your metabolism, these things become much easier.

However, the mid-Cretaceous thermal maximum didn't mean that the most warm-blooded dinosaurs went extinct; their diversity waned, but they persisted, and in fact adapted to the changes. In Gondwana, we see a new breed of theropod emerge called the abelisaurs. These predators are descendants of the more ancient *Ceratosaurus* who roamed North America in the Late Jurassic. While much of their evolutionary history has been mysterious between then and the Late Cretaceous, the abelisaurs were significantly different from their older relatives, and radically different from many of the other theropods at the time. In an increasingly feathered world, abelisaurs were quite archaic in being covered in scaly skin and bony scutes, giving them a very reptilian look. Many of their heads were adorned with boney protrusions or even an assortment of horns covered in thick keratin. Despite reaching nearly six meters (twenty feet) in length, these dinosaurs had arms less than the length of a human forelimb. While T. *rex* is often thought to have useless arms, the abelisaurs truly did. With finger bones the size of pebbles and practically no muscles in the arm for movement, it may have been impossible for these limbs to do anything. Yet they compensated for this with powerful leg and tail muscles, unlike most other theropods. This provided them with explosive speed perfect for quick pursuit.

Meanwhile, in North America, the continent began to go through significant changes in its geography. Tectonic forces began to push up the western region into mountain ranges. At the same time, the Midwest was stretched thin, like taffy, and the land began to sag. These extensive mountain ranges and basins combined with the increasing sea level due to climate change paved the way for the Gulf of Mexico to fill in these low-altitude areas and thus create a massive interior seaway. This Western Interior Seaway spanned all the way from Houston, Texas, to the Eskimo Lakes of Canada. The new seaway cut North America into two massive islands, Laramidia and Appalachia, and brought all-new patterns of cool currents to the interior of the continent. With mountain ranges to the west and coastal plains to the east, the more temperate Laramidian ecosystem became a haven for new and unexpected dinosaurs during the thermal maximum.

Sauropods and non-coelurosaurs were still present and on top, although not as common as they were in places like Patagonia. Yet within the shadows of this old empire, unique breeds from East Asia had already migrated over and integrated into the ecosystem. Before sea level reached its peak during the thermal maximum, the land bridge known as the Bering Strait, which connected Asia to North America, was still open. This Bering Strait facilitated an influx of new Asian dinosaurs into Laramidia that fit well into the new shifting environment.

By the time the thermal maximum hit, there lived many dinosaurs that looked awfully similar to those that lived in the snowy forests of the Early Cretaceous. Among them were small, parrot-beaked, thick-headed herbivores known as the ceratopsians and pachycephalosaurs. Others included offshoots of the iguanodonts known as the hadrosaurs. Species like *Eolambia* had specialized jaws for chewing, but also sported a large duck bill with extensive nostrils. Among the plethora of coelurosaurs that crossed over was a unique type of predator. While the proceratosaurs of Asia had gone extinct after the initial heat snap of the Early Cretaceous, a small remnant lived on. They were smaller, faster, yet still retained their powerful jaws for such a small animal. These were the tyrannosaurs. The ceratopsians, pachycephalosaurs, hadrosaurs, and tyrannosaurs might all sound familiar, as their descendants were some of the most iconic dinosaurs: *Triceratops*, *Pachycephalosaurus*, *Edmontosaurus*, and *T. rex*. During the global warming of the Middle Cretaceous, they were hardly as impressive as their later descendants. Many of them were puny in comparison. *Moros*, the earliest tyrannosaur in North America, which lived at the peak of global temperatures, was little more than the size of a wolf. Early ceratopsians would have only been about the size of wild boars. So, how did these dinosaurs go on to become so big and terrifying?

All of this extreme heat eventually subsided. The large igneous provinces that began erupting at the end of the Early Cretaceous had finally finished. As a result, there was a significant drop in the amount of carbon dioxide being pushed into the atmosphere. With carbon dioxide no longer being input into the atmosphere, carbon sinks finally had a chance to absorb it and thus cool the planet. The newly evolved flowering plants especially took advantage of this. Before the thermal maximum, flowering plants only represented a fraction of global plants; they were still quite new and had developed during a rather cool point in Earth's history and gone through a tumultuous bit of climate change. Now, the abundance of carbon dioxide gave flowering plants the resources they needed to expand into both trees and low-lying vegetation. In fact, they were so successful that cycads were almost entirely replaced by them. Plankton also began to rebound after the eruptions. Now that the extra nutrients from the volcanism had stopped, plankton populations could stabilize and stop forming algal blooms that choked the ocean. Even at technically lower amounts, these healthy plankton populations still helped absorb carbon dioxide and cool the planet further. From all of the carbon absorption, the planet cooled at roughly the same fast pace that it took to emit all this greenhouse gas. Dinosaurs went through another span of climate change that was equal in scale, yet opposite in direction.

The result was a minor extinction; dinosaurs across the globe that were adapted for a hot climate died off and were replaced by dinosaurs adapted for a cooler climate. Less than 75% of all life went extinct; therefore, it was not considered a mass extinction. However, that did not mean the effects were any less important. The reign of the carcharodontosaurs and spinosaurs ground to a halt, and they went completely extinct as far as we know. *Mapusaurus* and *Spinosaurus* were the last kings of these two long-lasting dynasties. The rebbachisaurs, which had proliferated across Gondwanan ecosystems, died off entirely. As flowering plants began to replace the ferns and horsetails they so crucially relied upon due to their low browsing posture, they were left with food that was too difficult to eat, thanks to their fragile, comb-like jaws and teeth. The only lineage of sauropods that managed to escape the clutches of extinction were the titanosaurs. While they never again reached their insane sizes during this thermal maximum, they went on to inhabit multiple continents. In their place, this new assortment of dinosaurs waiting in the wings took over.

We don't seem to think that climate change, especially our own situation, could be beneficial to life. Yet, as we have seen in other situations, like the climate disasters of the Late Triassic, early dinosaurs benefited from those changes while ancient archosaurs suffered immensely. As strange as it sounds, many modern animals will become more diverse and successful in the coming years as others go extinct, as they are predisposed to warmer climates. In East Asia, scientists have carefully studied reptiles, like Mongolian racerunners and Amur grass lizards, and found that they were evolving in beneficial ways. As their environment warmed, they grew faster, ran faster, and even had a stronger immune system thanks to a changing gut biome. In Central America and the US, scientists transplanted brown anole lizards into new warmer environments to see how they would react to rapid climate changes. Rather than struggle, these lizards actually evolved a higher heat tolerance and quickly adapted to the changes. Even some species of butterflies and moths show signs of diversifying; as temperatures increase, their ranges expand, they emerge from cocoons earlier, and they produce more generations in a given year. By all accounts, Darwin would say these animals are doing great. In the coming years, as temperatures increase and many warm-blooded species decline or even go extinct, we'll likely see a proliferation of cold-blooded animals. Reptiles, amphibians, insects, and other animals will become more common. Warm-blooded animals may even give up on being strictly warm-blooded and embrace a more mesothermic lifestyle. Climate change doesn't just affect the weather or the severity of storms; it affects all animals and whether they succeed or fail in nature.

Something like the faunal turnover of the Middle Cretaceous may be the likely future for us. We're already on this trend, as temperatures increase and species around the world go extinct. If we allow carbon dioxide emissions to continue for the next thousand years, we're estimated to hit 1,284 parts per million by 3025, which would increase global temperatures by 9 °C. Yet, if we do course-correct, a minor extinction will be the best-case scenario. That may not sound comforting, but a minor extinction is preferable to a mass extinction. It's better to accept the damage that has been done and take actions to fix it than to rack up higher biological and ecological debt in the future. Something to ease the anxiety is knowing that life will balance itself in situations like this. Yes, many dinosaur groups went extinct during this time, but also, many types of dinosaurs that were more cold-adapted made it through to the other side of this crisis. The Asian immigrant groups in North America, the abelisaurs of Gondwana, and even older types of dinosaurs like the ankylosaurs, the megaraptors, many other coelurosaurs, and of course birds, all made it through the thermal maximum. So too will modern animals handle our current situation. Warm-blooded animals like mammals and birds are expected to decline as temperatures increase. Yet with corrective action, this decline will subside and they will make a rebound as temperatures return to normal.

As stated earlier, returning to normal can't happen as quickly as we are warming the planet; even in a preferred direction, climate change must take time. If we allow emissions to continue until 3025, it will take roughly 218,000 years to return to our current temperatures. Yet, after that, if we continue to end carbon dioxide emissions, we will stay in a climate that is more or less optimized for humanity for at least a million years into the future. That climate optimization will stay that way as long as carbon dioxide levels remain around 250 parts per million. It's hard to comprehend such a large span of time. It may seem unattainable, or even futile to try, to effect change that far into the future. However, it isn't. Anatomically modern humans have existed for three hundred thousand years. We are the result of three hundred thousand years of decisions: granted, not complex decisions like this, but equally important to our survival. Our descendants, no matter how far into the future, will be thankful for the decisions we take to improve humanity one step at a time.

It seems clear that the cooling that reversed the Middle Cretaceous global warming was due to the rebound of carbon sinks; phytoplankton returned to a healthy abundance and plants, especially angiosperms, blossomed. Additionally, with the separation of the continents facilitating the circulation of cooling currents across the globe, the planet cooled much faster than when all land was conglomerated in Pangea. This is the optimistic news of our situation: The composition of the continents and the global distribution of plants has never made it easier to cool the planet and rebound from global warming. That being said, is our modern distribution of flora able to take on climate change? Do we have enough trees?

Over the past three hundred years, deforestation has resulted in the loss of nearly 50% of the world's forest area. Yet forest loss has not been evenly distributed; some countries have not lost much, while others have lost nearly all of their forests. It is estimated that twenty-nine countries have lost 90% of their forests. The tropics are especially vulnerable to deforestation, as they hold a majority of the world's forests and therefore convert much of the atmospheric carbon dioxide into oxygen. Through observations made by satellite imaging, Earth lost 7.6 to eight million hectares of forests from 1990 to 2010 in tropical regions. This is roughly the area of Scotland being lost to deforestation every year. Thus, the amount of carbon being absorbed from our atmosphere is decreasing. In the 1990s, an estimated 115 megatonnes (127 million tons) of carbon dioxide were absorbed by global forests per year, but only ninety-seven megatonnes (107 million tons) were absorbed per year in the 2000s. Just as ocean anoxia leaves patches of Earth's waters unlivable, deforestation is leaving some portions of our land unlivable for nature and incapable of converting carbon dioxide into oxygen.

The tropics are also home to one of the largest carbon sinks in the world: the Amazon rainforest. As one of the largest forests in the world, it is exceptionally potent in absorbing carbon dioxide. It currently holds roughly 150 billion metric tons of carbon dioxide, thanks to its sheer size and centuries of steady absorption. It's essentially the most efficient factory for turning carbon dioxide into the oxygen we need. However, deforestation threatens this. Nearly 219,889 kilometers2 (eighty-five thousand miles2) have been destroyed through deforestation. That's roughly 30% of the area that can no longer sequester carbon dioxide. At the same time, deforestation has led to the release of the stored carbon dioxide in the trees and soil. As the Amazon rainforest shrinks, its ability to absorb carbon dioxide and release oxygen decreases as well. Therefore, the Amazon represents an essential area to not only conserve, but hopefully restore through expansion as well. The Amazon is not the only place that is threatened in the tropics, though. Places like Southeast Asia and Africa also are experiencing deforestation. It's estimated that only 24% of tropical forests in these regions remain intact. Deforestation in the tropics has also led to many forests being fragmented; forests have been cut through and extensively bisected. This

means that forests are not one large, moderately stable patch
of land; instead, they are several small patches. The absorbing
and conversion effects of forests compound with size: Like
islands in the ocean, smaller forests can't convert as much
carbon dioxide and struggle to support ecosystems as life
becomes disconnected.

It seems clear that one of the most important ways to
curb global warming is to encourage reforestation to a point
where carbon sinks absorb more carbon dioxide than is being
emitted into the atmosphere, thereby creating a net decrease
in greenhouse gases. However, based on the timescales of the
past, this realistically can't happen overnight. It took over
one hundred million years to cool the planet after the end-
Permian extinction. As Pangea broke up and plants engulfed
the planet, it only took about thirty million years after the
mid-Cretaceous thermal maximum to cool the planet back to
levels similar to today. With the current configuration of the
continents and the influence of humans, we can bring that
timeline down significantly. We need to start small. Focus on
emitting less carbon dioxide than last year (rather than more,
as we have been trending). Make it a goal to cut down fewer
trees than last year. Only after we make progress toward
slowing things down can we realistically cut emissions to
zero or expand forest area.

As the planet warms and cools, life will adapt in
surprisingly hardy ways to ensure its survival. The dinosaurs
that survived the thermal maximum survived not just

because they were smaller and less specialized. All of them also had unique structures that allowed them to remain cool in such oppressive heat. In North America, despite a majority of the new dinosaurs originating in Asia, many were not closely related. Tyrannosaurs were about as distantly related to the herbivores as they could be. The armored ankylosaurs had split hundreds of millions of years earlier from the neornithischians. Even among the main neornithischians, the ceratopsians and hadrosaurs, there were very few similarities at this point. Surprisingly, all of these distantly related dinosaurs all had one thing in common: their noses. Their noses weren't literally all the same, but they convergently evolved a similar function of keeping the brain cool. The internal structure of the upper snout within tyrannosaurs, and indeed many other coelurosaurs and birds, expanded to create an enlarged nasal cavity. Ankylosaurs, as mentioned earlier, evolved their long and convoluted nostril passages millions of years before. Hadrosaurs, known for their iconic duck bills, sported nostrils as big as their eye sockets even in the earliest species. Even the smallest ceratopsians had relatively large nasal bones. With each breath, these unique noses functioned as efficient heat exchangers; warm air entered the nostrils, cooled on its journey through this cavity, and thus prevented the brain from experiencing heat stroke. This likely helped these warm-blooded dinosaurs persevere through the thermal maximum.

To survive the hot and dry conditions of Gondwana, abelisaurs relied on their most ancient dinosaurian feature: air sacs. Indeed, despite looking very different from birds, they had one of the most unique characteristics of birds as well. Like their distant bird cousins, abelisaurs had hollowed-out vertebrae and ribs that made space for large air sacs under their necks and in their hips. Combined with their lungs, their entire respiratory system covered nearly a third of their length. In early species that lived during this time, like *Skorpiovenator*, this respiratory system may have reached up to 1.25 meters (four feet) in length. This provided these animals with an expansive internal air conditioning system. Their thick skin also extended from the head down the neck, to their shoulders, flanks, and the base of the tail. While it was great at providing protection from rivals, it also simply protected these animals from the dusty air that would strip away fragile layers of skin. Similar to modern animals, this thick skin made it harder for moisture to escape through the skin and helped retain water in these arid environments. Their massive air sacs and thick scaly skin seem to have been effective, as abelisaurs were barely ever found outside of hot and dry climates. Titanosaurs obviously benefited from the extreme heat due to their poikilothermic metabolism. Yet they still managed to keep themselves cool, thanks to their porous bones and air sacs, as well. Not only did their pneumatic bones make them lighter overall, it also allowed their air sacs to cover more of their body and keep them cool. As descendants of the macronarians—the large-nosed sauropods—they too had large nostrils to help cool their heads. Even though they survived the minor extinction, they evolved much smaller body sizes (relatively) to conserve metabolic energy.

For the remainder of the Late Cretaceous, roughly twenty million years, Earth began to recover from the intense heat. There were occasional brief periods of punctuated cooling events, but for the most part it was a gradual trend of colder global temperatures (although it was still a much hotter world than today). Even at its coldest, global temperatures likely never dropped below 20 °C (68 °F). Never again did dinosaurs see permanent ice caps in the polar regions. As the planet cooled, it transformed from a strictly uniform, global, tropical climate to unique biomes in each continent. Each region slowly developed its own schedule of seasons and temperature ranges. These unique biomes became more diverse as time went on, until within each continent there existed a surprising amount of regional diversity. So diverse, in fact, that in many cases, regions of one continent overlapped with another.

In theory, a dinosaur group in one region of South America could live just as well in a certain region in Europe. This isn't just theoretical; the fossil record lines up with it. Across the globe, we find dinosaur groups whose lineages can be traced back to completely different continents. In the island chains of Romania, we find sauropods whose close ancestors are derived from South America, Africa, and Asia. In France, abelisaurs strangely pop up, despite being found in the opposite hemisphere. Tyrannosaurs and ceratopsians hop back and forth between western North America and China and Mongolia. Hadrosaurs miraculously find a way to cross the Mediterranean from Iberia to Morocco. Paleontology has reached a point where we have enough evidence to realize that the migratory patterns of dinosaurs in the Late Cretaceous were exceedingly complex. However, we don't have enough evidence yet to parse through the complexity and confusion to nail down the exact routes and timing of these migratory patterns. This becomes stranger given that sea level was quite high during this time. There weren't grand open patches of land connecting these continents, at least for a measurably long time. Explanations range through many hypotheses. Some posit quick bursts of migration as sea levels briefly drop. More extraordinary claims suggest that, after massive tropical storms, rafts of plant debris jumbled together into giant rafts holding dinosaurs. These rafts floated across seas and giant rivers to other land masses or islands. Such a hypothesis seems fantastical, yet things like this happen today. Monkeys, birds, and iguanas in tropical regions have been swept up in storms and sailed off to nearby islands. Regardless of the explanation, the effect was that dinosaurs began to go through another stretch of vicariance evolution.

By the end of the Cretaceous, the dinosaur remnants from the faunal turnover had flourished into a diversity of unique and iconic dinosaurs. Along the estuaries of Patagonia, the unique Gondwanan fauna asserted their dominance. Temperatures were relatively cold, and reached down to freezing winters especially in the southernmost tip, similar to today. The radical change in climate from the Middle Cretaceous spawned an equally radical change in the herbivores that grazed this region. Instead of titanosaurs comfortably reaching over thirty meters (one hundred feet) in length, the titanosaurs of this region were practically pygmies by comparison. Species like *Titanomachya* (whose name even pays homage to the end of the Greek Titans' rule) reached six meters (twenty feet) in length and weighed only 7.8 tonnes (8.6 tons). Yet even at their smallest, they rivaled even the biggest of modern animals. The unique metabolism of sauropods meant they had to shrink in size to accommodate living in an increasingly cold world. With size no longer an overwhelming deterrent to predators, titanosaurs relied heavily on these osteoderms for protection. Yet despite being significantly smaller, this actually made them much faster. While *Titanomachya* was a tenth the size of the behemoth *Argentinosaurus*, it was estimated to run twice as fast. These defenses and their speed were indeed put to the test by the ferocious predators that hunted them.

As a connection point between the mainland of South America and Antarctica, Patagonia saw the confluence of two worlds of predators. The abelisaurs had grown significantly up to this point and hardened their heads to sport intimidating horns or gnarled snouts. *Carnotaurus* was the largest in the region, reaching eight meters (twenty-six feet) in length, but other smaller abelisaurs lived alongside it as well. With such similar body types, competition between these predators might have spilled over. Yet their different sizes likely kept them within their own lanes; large species like *Carnotaurus* could bring down larger prey, possibly titanosaurs, while smaller species like *Koleken* used their explosive speed to catch smaller, nimble prey. Yet only a few hundred miles south was an equally dangerous king whose lineage hailed from the South Pole of the Early Cretaceous. *Maip*, the largest megaraptor to have existed, reveals that these polar dinosaurs survived the harsh warming of the Middle Cretaceous and expanded their domain. With arms easily over one meter (three feet) long, complete with deadly hooked claws, they were completely different from the short-armed abelisaurs. As such, they still likely snatched up small ornithopods as prey, but standing over ten meters (thirty-three feet), they may have entertained other options. Even below the massive megapredators of South America there existed a special diversity of unique theropods: small and nimble cousins of the abelisaurs called the noasaurs that used their strange outward-curling teeth to catch invertebrates. Massive toothed birds known as the unenlagiines used their pointy snouts and thick teeth to pluck fish from streams. Fauna like this ruled South America, but their dominion spanned all across the other former Gondwana continents.

Africa and the islands of Madagascar and India were mostly hot and arid equatorial regions, making them perfect environments for abelisaurs and titanosaurs to expand into.

Across the equator existed an entirely different world. The ecosystem of North America had very little in common with South America, suggesting that dinosaurs rarely crossed over into either continent. However, it was no less impressive. Much like today, North America featured a unique spread of ecosystems. From the warm and dry coastlines of Mexico to the redwood-filled forests of Montana, and all the way up to the frozen boreal forests of Alaska, there were plenty of environments for dinosaurs to carve out their niche. At the same time, the growing range of the Rocky Mountains and the shifting coasts of the Western Interior Seaway created barriers all across Laramidia. With dinosaur populations cut off from each other and regional climates differentiating, the western half of North America exploded with diversity.

Chief among them were the tyrannosaurs. The small coyote-sized predators of the Middle Cretaceous quickly evolved to the size of Clydesdale horses, then to larger than all modern predators, and finally reached the pinnacle of monstrous size which dwarfed even elephants. This pinnacle was the infamous T. rex. At over twelve meters (forty feet) long and weighing over eight thousand kilograms (seventeen thousand pounds), it was the heaviest predator ever to walk the Earth. Putting its behemoth size aside, tyrannosaurs were ecologically some of the most terrifying predators ever. Their warm-blooded metabolism allowed them to grow at rates unlike most animals back then, and even today. A T. rex could reach full adult size in roughly thirty years—that's an average of 267 kilograms (567 pounds) gained every year. This astronomical growth rate and need for food allowed it to fill different niches as it grew. As long-legged juveniles, tyrannosaurs were the fastest predators in their ecosystem and could chase down small feathered dinosaurs. As they grew and slowed down, they turned into highly efficient walkers. Whether T. rex was fast enough to ambush prey is up for debate, but regardless, it would have been capable of stalking its prey to death through endurance hunting, much like modern wolves. By occupying different niches at different ages, tyrannosaurs essentially filled in spots that could have been filled by other smaller to mid-size predators—a very different dynamic to the Serengeti of the Late Jurassic. Tyrannosaurs were also not seasonal predators. Despite its moderately warm summers, North America was one of the coldest regions on Earth, especially in places like Alaska and northern Canada. The cold weather didn't seem to bother tyrannosaurs; as relatives of Yutyrannus, tyrannosaurs were likely covered in feathers that kept them warm in winter. No need to migrate based on the seasons. They had a near monopoly on their ecosystem.

It wasn't entirely a monopoly, though. North America featured a variety of raptors that hunted small prey. In

North America, dromaeosaurs ranged from two-meter-long (six-foot-long) hunters, like *Acheroraptor*, to even grizzly-bear-sized predators in the north. By the End Cretaceous, raptors had become a staple across most environments. While tyrannosaurs stuck to comparatively colder environments, raptors had a broader distribution thanks to their smaller body size; at smaller sizes, raptors were less prone to overheating than tyrannosaurs and thus could be found much closer to the equator. This was also the case for the many other herbivorous coelurosaurs of North America. The special combination of feathers and endothermy for warmth and air sacs and hollow bones for cooling meant that they had the innate ability to inhabit just about any environment. Across the entire continent, we find an array of unique coelurosaurs coming into their prime. Dinosaurs like *Struthiomimus* (which literally translates to "ostrich mimic") had begun to diversify as some of the fastest dinosaurs. Rather than stand and fight against the quick and agile tyrannosaurs or raptors, these dinosaurs convergently evolved a body similar to modern-day ostriches, their long legs allowing them to reach speeds upwards of sixty-one kilometers per hour (thirty-eight miles per hour).

Yet not all herbivores were coelurosaurs. Titanosaurs were a rare breed here, with only one species living in the American Southwest. Instead, ornithischians ruled. Hadrosaurs had blossomed across Laramidia as they accentuated their nasal features. Some species featured massive, bony noses while others sported extravagant, hollow head crests that connected to their nostrils. These head crests probably provided some cooling benefits, but they also could produce loud trumpeting sounds. Each crest had its own shape, which likely let out its own style of deep honk to communicate between individuals or signal to the herd. Ceratopsians grew in variety as well. No longer cat- to pig-sized animals, dinosaurs like *Triceratops* acquired their own extravagant headgear in the form of horns and frills. While it's always been thought that these were used as defense, that may have been the case for only a few species. The horns of many of these species pointed all over the place, and the frills nearly always had massive holes in the centers that would have been covered by a membrane of skin. Instead, they may have provided multiple functions, ranging from sexual selection to predator intimidation. Given that they lived in the aftermath of global warming, it's also been thought that the large surface area of these frills effectively dissipated heat.

It's amazing and hard not to notice that the features that helped keep these dinosaurs from overheating likely had a dual purpose in sexual selection; it seems the ability to regulate body temperature and cope with the climate was an attractive feature. Life wants to survive; a single organism can't live forever, but it can make the next generation better. So too will life in the face of our current climate change. All the species on this planet will not roll over and die. Through sexual selection, they will choose mates that they think have genes that will benefit future generations. Those who are wrong go extinct. Those who are right get their offspring to keep playing the game of life. Yes, a minor extinction is at our doorstep, and yes, we should be doing everything we can to prevent it. While you're reading this, though, nature is unconsciously working hard too, in order to preserve its genes. Not all life makes it through extinctions, but some life does. Some life is resilient and best fit to handle whatever this planet throws at it and take over the niches left behind. Yet it's not up to us to have an entire planet of plants and animals conform. Neither should we remove ourselves completely from the equation. A balance needs to be struck—an equilibrium as the end goal—which is what organisms want in the end, too.

This dynamic of dinosaurs was not exclusive to North America. Many of these major groups were also found in East Asia. Tyrannosaurs, raptors, coelurosaurs, birds, ceratopsians, hadrosaurs, and ankylosaurs had crossed the Bering Strait many times throughout the Late Cretaceous, and so had established themselves in both continents. Minor differences existed, but overall eastern Asia and western North America had the same fauna. This is because the environment of East Asia was similar to North America in several ways. The regional climate was overall quite cool and semi-arid. It featured mountain ranges and valleys. Yet it was different in that the temperate forests of East Asia had given way to an expansive desert. While home to many dinosaur groups that lived in North America, the climate of East Asia was still warm enough in some regions to support some giant sauropods. Winters were quite cold, but summers were actually some of the hottest on Earth, roughly 30 °C (86 °F), allowing sauropods to survive. Here, sauropods could reach roughly twelve meters (forty feet) long. With the domination of ornithischians as mid-sized herbivores, only sauropods that exclusively ate from the tree canopies avoided competition with the plethora of other herbivores. As East Asia cooled from the thermal maximum, it saw the rise of a dinosaur exceptionally similar to modern birds.

These were the oviraptorosaurs, ostrich- to turkey-sized theropods that were covered in feathers and featured a toothless beak. Nearly all species were adorned with a semi-circular crest, similar to the modern-day cassowary. These were not new dinosaurs; the earliest species known so far have been found in the icy Chinese forests of the Early Cretaceous. During this time, they were relatively small and obscure coelurosaurs—that is, until the Late Cretaceous saw these dinosaurs spread across the region. One feature of the oviraptorosaurs that stood out was that they were some of the first dinosaurs to roost over their eggs. Nearly all dinosaurs would simply bury their egg clutches underground, like modern-day crocodiles, to keep them warm while they grew. This keeps the eggs safe and hidden while also freeing up time for the parents to hunt or gather resources. The tradeoff is that the environment has full control over whether the clutch survives or not. If temperatures change too much, then the clutch doesn't survive. Additionally, as reptiles, the temperature of the egg actually determines the sex of the baby; the hotter it gets, the more females in a clutch, and the colder it gets, the more males in a clutch. Oviraptorosaurs took matters into their own hands. Instead, these dinosaurs would keep the nest open and sit over their eggs. While this required more work on the parent's side, it allowed them to actively monitor the health of the nest; they could incubate them if it got too cold, or let them air if it got too hot. Early paleontologists thought many of these dinosaurs died while stealing eggs (the first genus to be discovered, *Oviraptor*, translates to "egg thief"), but more complete fossils reveal that it's quite the opposite. These dinosaurs commonly died huddled over their nests, sometimes buried by sandstorms. These exceptional parents would rather die with their clutch in the hope of protecting them than forsake them to the elements. This feature was clearly successful, as it not only allowed them to weather sixty million years of climate change during the Cretaceous; it was a behavior passed down to their close relatives, raptors and birds. Thus, dinosaurs that cared for their eggs can be found across every single continent. Whether it's in the poles or the equator, barren desert or humid and tropical, parental care over their eggs was a behavior in dinosaurs that could be found in every single type of environment.

All of this dinosaur diversity seemed to converge in Europe. Here, we see an influx of sauropods with lineages from all over the world. From Asia, Africa, and all the way from South America, something about the climate in Europe made it valuable enough for sauropods to cross the entire planet for it. Interestingly, Europe didn't have a diverse climate, but it did overlap with many regions of Asia, Africa, and South America. It had mild summers and warm winters—something sauropods likely relied upon for their metabolism. While the regional climate would usually allow these dinosaurs to become behemoths, they shrank to their smallest sizes in Europe. Many species, like the Romanian *Magyarosaurus*, reached only three meters (ten feet) long. In fact, they were so small that their young would have been easy pickings for the giraffe-sized pterosaur known as *Hatzegopteryx*. The European archipelago forced these sauropods to develop a condition known as insular dwarfism. In island ecosystems, herbivores commonly shrink because of the lack of space and resources to support massive sizes. This happens very quickly; gigantic species can evolve to pygmy sizes thirty times faster than pygmy animals evolving to gigantic sizes.

Titanosaurs weren't the only herbivores to develop this feature. The more warm-blooded herbivores, ankylosaurs, ornithopods, and even hadrosaurs, shrank down to the sizes of modern animals. Like the titanosaurs, all of the plant-eating species of this time period had lineages that could be traced across the globe. While the islands of Europe forced dinosaurs to shrink in size, it seemed a valuable tradeoff in order to live in a moderate climate fit for a diverse array of dinosaurs. It wasn't just preferable to herbivores; theropods of all kinds that would never meet in other circumstances gathered here. The abelisaurs of the southern hemisphere converged with the coelurosaurs prominent in the northern hemisphere. The warm island climate was not ideal for tyrannosaurs, so abelisaurs assumed the role of apex predators. However, it became a haven for coelurosaurs, and especially birds.

The coastlines of Europe, and indeed the coastlines of the globe, featured an assortment of unique shorebirds. Ancient shores are an abundant source of fossils as the tides pull the corpses of dinosaurs and birds out into the ocean where they are buried in the seafloor. Other times, flowing rivers and floodplains quickly bury these birds. At the end of the Cretaceous, we see a plethora of ancient birds distantly related to birds of today. Despite the lack of relation, many species convergently evolved bodies awfully similar to birds of today. In North America, toothed birds evolved into eagle-like predators in the inner forests, while evolving seagull-style bodies close to the shore. Long-beaked toucan imitators inhabited the northern regions of Madagascar. Even in the tropical island of Romania, giant flightless birds roamed the land. While non-avian dinosaurs took center stage as the biggest, fastest, most ferocious animals on Earth, the success of prehistoric birds at least equaled if not surpassed them.

Among all these prehistoric birds lived a few species that are truly special. In tropical Belgium, we find an overlooked bird that had profound consequences for the future of dinosaurs. Weighing 394 grams (13.9 ounces) was *Asteriornis*, one of the earliest ancestors of modern birds like chickens, turkeys, and quail. Like all modern birds, *Asteriornis* had a toothless beak, a fingerless wing, and no tail. Just like its descendants, the game fowl, it had a combination of features that amounted to a thin beak and roof of mouth (or palate). All this is to say, *Asteriornis* is undeniable evidence that "modern" birds existed alongside dinosaurs. Modern birds weren't restricted to the island chains of Europe. All the way on the other side of the world, in frigid Antarctica, we find *Vegavis*, a small flying bird that is an ancestor to birds like ducks, geese, and swans. These two birds suggest that modern birds weren't just exotic dinosaurs in far-off remote regions; they were ubiquitous across the globe.

Being able to survive in two radically different environments suggests their metabolism was versatile enough to handle all sorts of climates. Their hollow bones and air sacs helped them survive warm environments, and their ability to generate heat and trap it through their feathers allowed them to brave even polar conditions. Their generalist metabolism was complemented by a generalist diet. The beak of *Asteriornis* was about as basic as it could be; no specialized shape, no curved features like those of birds of prey, and no robust beaks of those found in seed-eating birds. Much like their distant silesaur ancestors in the Triassic, modern birds would have been jacks-of-all-trades. It was crucial to have a generalist diet, as they were not only the smallest of the dinosaurs, but some of the smallest birds among their prehistoric toothed and clawed relatives. It was a tough life for modern birds at the end of the Cretaceous, but it was about to get even tougher.

Far in the outer reaches of space, originating beyond the asteroid belt by Jupiter, something had been approaching Earth. Traveling at nearly seventy-two thousand kilometers per hour (forty-five thousand miles per hour), fifty-eight times the speed of sound, was an asteroid twenty-one kilometers (thirteen miles) in diameter. It's unknown what set this city-sized asteroid on a path toward Earth, but what is certain is what happened when it collided with Earth sixty-six million years ago.

It was the end of the dinosaurs.

When this asteroid hit the Yucatán Peninsula in modern-day Mexico, the impact created an explosion with the force of seventy-two trillion tons of TNT. It was several times more destructive than the Manicouagan and Carswell impactors that came before. It produced a crater 150 kilometers (ninety-three miles) in diameter, and likely vaporized all life in Mexico in an instant. What followed this explosion put even Hollywood disaster movies to shame. A blast of air rushed out from the explosion at 1,046 kilometers per hour (620 miles per hour), flattening forests and even the biggest dinosaurs alike. The ensuing shockwave caused global earthquakes ranging in magnitude from nine to eleven. So intense were these shockwaves that they barreled through the surrounding gulf and sent one-hundred-meter-high (330-foot-high) mega-tsunamis crashing into the surrounding landmasses. Marine life was completely ripped apart and thrown all the way into the interior of Texas and Florida, thousands of miles from the impact site. Without a doubt, much of North and South American life was wiped from the face of the Earth within minutes. T. rex and Triceratops—some of the most powerful dinosaurs to ever walk the Earth—were promptly eradicated. The rest of the world was not safe either. The epicenter of the explosion was so hot that it melted three thousand megatons of the Earth's crust. This molten rock was ejected into the atmosphere, and coagulated into untold amounts of pebble-sized molten spheres called tektites, which then rained down on the Earth. Dinosaurs across the planet were pelted with these molten pebbles. Only animals small enough to hide were safe. White-hot rocks whizzed through forests and they erupted in flames. Wildfires spread across every continent. The world burned.

Not all of the ejected debris made its way back to the surface of the Earth. Like the Manicouagan impactor, much of this debris still hung in the stratosphere. Yet this time, the sheer scale of debris was able to blot out the sun across most of the planet. The world was plunged into a frigid and unending night. For months, Earth plummeted 7-12 °C into a bitter cold snap nearly as cold as our most recent ice age. Even with a vast majority of dinosaurs being warm-blooded and even feathered, the change was too much and too quick. With no light and no warmth, the final event in the chain of extinction unfolded: Plants withered and died. As a result, herbivores starved to death, starting with the biggest dinosaurs down to the smallest ones. Finally, the predatory dinosaurs cannibalized themselves and then starved as well. By the time the sky cleared, Earth's life forms had been massacred. Pterosaurs, mosasaurs, plesiosaurs, and hundreds of other species succumbed to extinction, but, most importantly, the dinosaurs were gone. It was the second-worst mass extinction in Earth's history, second only to the End Permian. From the ashes of those global volcanoes in the Permian, the dinosaurs rose to rule the world. In the ashes of the impact winter in the Cretaceous, their dynasty crumbled and was never seen again.

Or was it?

CONCLUSION

Altered Place

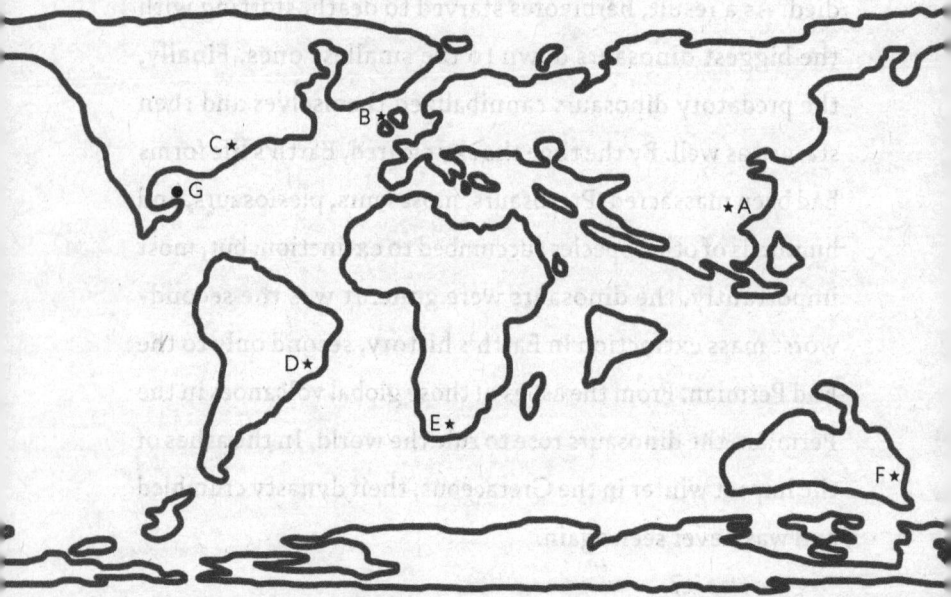

Pangea: ~66 million years ago

A - Beijing	C - New York City	E - Johannesburg	G - Chicxulub
B - London	D - Rio de Janeiro	F - Sydney	impactor

In the ashes of a global impact winter, it seemed as though all hope was lost. With life eviscerated by the biggest asteroid impact in Earth's history and the climate plunged instantly into a cold snap, it may appear as if it would have been impossible for life to recover. Yet life always seems to find a way.

In the Cretaceous, less than one million years before the asteroid impacted, a volcanic province known as the Deccan Traps began to crop up on the island of India. Coalescing into a crater over fifty kilometers (thirty miles) in diameter, this volcano began to pour out lava all across India. Over 1.5 million kilometers2 (0.58 million miles2) was flooded in lava. From roughly Goa to Madhya Pradesh—half of the area of India—environments were engulfed. Abelisaurs, like *Rajasaurus*, and titanosaurs, like *Isisaurus*, were at the epicenter of what seemed to be another period of intense global warming. It was the intense eruptions of the Deccan Traps that were originally thought to be what killed the dinosaurs. Volcanoes of this scale were the most likely culprit for extinctions, as we've seen. That is, until the ash layer above the last dinosaur fossils was more closely examined and was found to contain things like shocked quartz, microtektites, and iridium levels so high they could only have come from an asteroid impact. Instead of killing off the dinosaurs, the Deccan Traps may have actually saved life from total extinction.

Due to these eruptions and their emissions of carbon dioxide, Earth was already starting to warm just before the impact. This actually softened the blow of the impact winter; without the volcanism, it likely would have gotten much colder. The asteroid impact did not stop the volcanic eruptions of the Deccan Traps, either. Carbon dioxide continued to be emitted during the impact winter and thus began to warm the planet immediately after the cataclysm. Without the Deccan Traps, Earth might have been stuck in sub-freezing conditions for three to sixteen years and wouldn't have reached pre-impact temperatures for thirty years. However, thanks to global warming, it likely took less time for Earth to recover.

From the ashes of the devastation rose a diverse group of survivors on land that are still here today: arthropods, insects, amphibians, lizards, snakes, turtles, and crocodilians, to name a few major groups. However, among the post-apocalyptic life were two special types of animals, the first being those distant descendants of the therapsids that ruled the Permian: mammals. Yes, by the end of the Cretaceous, our shrew-sized ancestors had evolved, albeit in the shadow of the behemoth dinosaurs. In a reversal of the last 180 million years, mammals would rise in place of the dinosaurs and take over the planet. They would spawn a unique diversity of animals, including humans, that adapted to this new world without dinosaurs.

But it actually *wasn't* a world without dinosaurs. One unique and unorthodox dinosaur was able to survive the destruction: birds. Not just any type of bird, though. Many birds were decimated by the impact. More ancient types that still had teeth, claws, and/or tails, were wiped out. Only the descendants of *Asteriornis* in Belgium and *Vegavis* in Antarctica lived on to proliferate into the overwhelming diversity of birds we see today. In fact, the diversity of modern birds dwarfs the known diversity of birds before the impact. Why? Why did this one specific type of dinosaur manage to survive one of the worst extinctions in history, but none of the other countless species of non-avian dinosaurs could brave the impact winter?

If we look at *Asteriornis* and compare it to the very first dinosaurs that survived the end-Permian extinction, they share a suite of characteristics that make up the Lilliput effect and make them preadapted for disaster. They were very small, so they didn't need lots of food to survive in a resource-deprived situation. Any *T. rex* and *Triceratops* that managed to survive the impact would have been the first dinosaurs to bite the dust due to their massive size. Titanosaurs, too, were especially susceptible, as they relied on sheer plant volume, unlike all other dinosaurs, to stay alive. Birds like *Asteriornis* or *Vegavis*, at this point, were very similar in size to the animals that left the strange *Prorotodactylus* footprints at the beginning of the Triassic. The small size of birds also meant they could hide from the hellfire and destruction while the bigger dinosaurs were stuck in the open, unprotected.

Birds were also generalists and could eat anything. By the end of the Cretaceous, many dinosaurs had developed more specific diets. Hadrosaurs and ceratopsians, for example, ate different plants. Even different-sized tyrannosaurs or abelisaurs preferred different prey. This avoided competition between different species and allowed them to diversify in more specific niches. The catch-22 was that food webs were much more fragile when it came to disasters. Like the picky-eater therapsids, most dinosaurs at this point ate particular things. Once those food sources went extinct, the dinosaurs that ate them were quick to follow. Not birds, though; the basic beak shape of Asteriornis prevented it from diversifying into a specific niche, but that meant it could eat just about anything. Like silesaurs, they could eat nearly any plant, and even hunt other small animal survivors. In the same way that early dinosaurs were able to survive in the wake of the end-Permian extinction, so too were birds able to survive the end-Cretaceous extinction. History seems to repeat itself.

Since the changing climate and ecosystems slowly influenced the evolution of dinosaurs, birds had also acquired many other features that predisposed them to be extinction survivors. One key feature was the warm-blooded metabolism. As endotherms, they were able to generate their own body heat. Moreover, as homeotherms, they were able to maintain a constant body temperature, allowing them to warm themselves up if they were colder or cool themselves if they were hotter (in part thanks to their extensive air sacs). Their extremely fast growth rate also meant they spent less time as vulnerable juveniles dependent on the care of their parents. Modern birds can reach maturity in a year or less, and thus quickly become self-sufficient. Another was the ability to fly. Being highly mobile is crucial to seeking out better environments in an extinction landscape, and to catching increasingly small prey. Flight makes travel, hunting, and even escape much easier, with the additional bonus of being able to hide in elevated spaces, away from the majority of other animals. Nesting, too, was an advantage compared to other dinosaurs who didn't. Dinosaurs that didn't incubate their eggs themselves had their eggs freeze, while dinosaurs that did nest were able keep their eggs warm enough through the brief but harsh impact winter.

One thing that modern birds, like *Asteriornis* and *Vegavis*, shared that other prehistoric birds didn't have was their unique arrangement of feathers. Obviously, in order to fly, birds developed their unique feathers apart from the assortment of other proto-feathers that nearly all other dinosaurs had. Rather than just hair-like structures, the central rachis had vanes flanking each side. The soft down feathers would eventually be molted into the more rigid and aerodynamic flight feathers. Something modern birds retained that other birds seemed to lose was the soft down feathers that covered their young. Many other prehistoric birds instead quickly molted these baby feathers and grew flight feathers. While less insulated, these prehistoric birds began flying much earlier in their lifespan. In the Late Cretaceous, when the world was still quite hot just after the mid-Cretaceous thermal maximum, the tradeoff seemed to be worth it, as modern birds were not as successful as these other prehistoric birds. However, once the impact winter engulfed the planet, staying warmer became a more important priority than getting in the air earlier. Thus, many types of prehistoric birds went extinct along with the non-avian dinosaurs.

Of course, there may be many other features that likely helped birds succeed, as opposed to all other non-avian dinosaurs and prehistoric birds; this is by no means an exhaustive list, especially considering how much of our knowledge is based on the fragmentary remains of the fossil record. Nevertheless, these features reveal how 180 million years of climate change nudged the evolution of dinosaurs to produce birds who could not only live in every environment on Earth, but also survive a mass extinction brought on by a climate crisis.

Now, sixty-six million years after the end-Cretaceous extinction, we are the ones facing a climate crisis. The burning of fossil fuels is releasing literally tons of carbon dioxide into the atmosphere, similar in magnitude to the countless volcanic eruptions the dinosaurs faced, eruptions that spanned entire provinces and lasted for hundreds of thousands, if not millions, of years. In a relatively short time, we could emit enough carbon dioxide to reach levels that haven't been seen since the time of the dinosaurs—all of which could create global temperatures the likes of which were experienced during the mid-Cretaceous thermal maximum, the end-Triassic extinction, or even the end-Permian extinction. Not only that, but we are also on course to experience the many other side effects of a warmer planet.

All of the global warming the dinosaurs experienced came with an expansion of deserts. While the splitting of the continents will keep us from experiencing the endless deserts of Pangea, we aren't immune to an expansion of deserts. Even now we can see it: The Sonoran and Mojave Deserts are growing and running out of water. Across the world, in places like the Gobi Desert, the Sahara Desert, and the African interior, deserts are expanding and reducing how much hospitable land there is. Increasing temperatures also intensify the water cycle and cause air to retain more moisture. Like the Carnian pluvial episode brought on by increased heat, all of this leads to not just an increased number of storms but an increase in storm intensity. Even now, the Atlantic hurricane seasons are producing more storms and hurricanes are becoming more intense each year. As deserts expand, our land surfaces will become less able to absorb the precipitation and flooding will become more common and more extreme.

The effects aren't limited to land; our oceans and coasts are on track to change severely. Warmer oceans mean dissolved oxygen is expected to exit the ocean and create widespread anoxia. The increasing storms erode more of the surface, which sends more nitrate-based nutrients and compounds into the ocean for plankton to take in. Like in the Early Cretaceous, algal blooms are likely to form and exacerbate ocean anoxia. Just like the Early Cretaceous, our glaciers are likely to melt quickly, especially those at the poles. This arctic amplification means that the poles are significantly more vulnerable to climate change than many other regions on Earth. It also means this glacier melt is a significant contributor to global warming, given that an expanding ocean absorbs more heat.

With more ocean water, our sea levels will rise. Things of the distant past, like the Western Interior Seaway and the European island chains, may be in our future.

As we've seen time and time again, throughout the reign of the dinosaurs, extinctions were quick to follow abrupt changes in the global climate. Global warming or global cooling, it didn't matter. What mattered was the magnitude of change in the climate and how fast that change happened. It's estimated that by 2100, Earth will have likely warmed two degrees Celsius from what temperatures were like before the Industrial Revolution. Two degrees Celsius over the course of three to four hundred years is equivalent to some of the worst extinctions during the Mesozoic. Therefore, the rapid change in temperature puts us on the precipice of the sixth mass extinction: the Anthropocene extinction.

As with every mass extinction, the bigger and more specialized an animal is, the more likely it is to go extinct. What few megafauna we have—elephants, rhinos, giraffes, hippos—are the biggest targets for extinction. Even stranger animals, like anteaters, whose diet is extremely specific, are slated for extinction as well. Yet it's not just the animals that we can recognize or think are cute. According to the IUCN (International Union for Conservation of Nature) Red List of Threatened Species, 9,694 species of plants, animals, and fungi are considered vulnerable to extinction. 5,220 are considered endangered. 4,574 are considered extremely endangered. Seventy-seven are already considered functionally extinct in the wild. If temperatures continue to increase and all these species go extinct, that's nearly twenty thousand species. Only life that is already primed to live in hot and arid conditions will make it through this period of global warming. Xeromorphic plants with thinner and coarser leaves will dot the landscape. Animals that are ectothermic, poikilothermic, or both will become more prevalent, as strictly warm-blooded animals overheat. Our whole ecosystems and how they interact will radically change.

Even putting all that aside, this is ultimately a problem for humans. We rely on many of these organisms to keep our ecosystems healthy, especially the ones that surround the places where we live. These organisms are essential for keeping our society functioning. We rely on pollinators to help our crops grow our food. We rely on predators to keep pests down. We rely on livestock and game for food. With increasing temperatures and expanding deserts, how do we expect crops to continue to grow and be healthy? Our societies don't live in a vacuum; all major cities and empires of the past existed in fertile and healthy land. Even at the most fundamental level, we as humans are not adapted for deserts or tropics. We are meant to live in temperate forests and always have. As temperate forests migrate toward the poles or disappear altogether, crops and livestock will be harder to grow. Major cities that used to support thousands to millions of people will be abandoned as people migrate toward the poles to avoid the hot, arid, and highly seasonal climate.

What's more, our lungs are like those of the therapsids, not of the archosaurs. Our technically wasteful breathing is not adaptive to an atmosphere that is growing in carbon dioxide. We may not suffocate to death, but these higher carbon dioxide levels will significantly impact our health. Studies have shown that living at one thousand parts per million of carbon dioxide affects our brains' decision-making abilities and how well we can concentrate. Considering that we are on track to hit 1,284 parts per million by 3025, this is extremely concerning. At 2,500 parts per million—the low estimate for the mid-Cretaceous thermal maximum—cognition plummets as people struggle to accomplish simple tasks. Global warming and increasing carbon emissions are not just a climate crisis, an environmental crisis, or an economic crisis; it is very much a human and public health crisis.

It all seems like a grim future.

But it doesn't have to be.

Throughout the reign of the dinosaurs, these radical changes in global climate weren't forever, and they always corrected themselves in the end. That's because the technology to combat global warming was always there and was able to work even without human intervention: plants, trees, soil, and phytoplankton. These things naturally take in the carbon in our environment, whether it's generated by animals or by disasters, and transform it into breathable air for us and make the climate cooler at the same time. It's been this way ever since the first plants appeared, hundreds of millions of years before the first dinosaur showed up. Looking at the Mesozoic, we can see that periods of global warming were followed by a diversification of plants. After the end-Triassic extinction event, for example, we see the rise of conifer and cycad trees. After the mid-Cretaceous thermal maximum, flowering plants exploded in diversity. Both of these situations set the planet on a course for global cooling.

So, where does this leave us? How do we move forward with this knowledge of ancient climate change? I'm not naive enough to think science can easily translate into exact policy. If anything, figuring out why we have a problem is way easier than providing an answer. By looking at climate change in the past and how it rectified itself, we have some broad guiding principles.

First and foremost is that the use of fossil fuels should *eventually* end. Not today. Not tomorrow. Not this year. Somewhere between "as soon as possible" and "as feasible as possible." We're stuck between a rock and a hard place, given that these fuels are the bedrock of our societies, yet burning them produces carbon dioxide that is rapidly warming our planet. Before we gut all infrastructure, we should start by focusing on not emitting more carbon dioxide than the previous year, especially considering that carbon dioxide emissions have been steadily increasing each year. Once we can accomplish that, then we can focus on decreasing emissions, and then ending them altogether.

Another important thing to note is that fossil fuels are destined to be a thing of the past anyway. Coal, oil, and natural gas are nonrenewable resources; since they are minerals and inorganic compounds, they are the byproducts of geochemical processes. These geochemical processes take millions of years to create these resources. We use them faster than they are being naturally produced. So eventually, maybe even in the near future, fossil fuels will become increasingly scarce—to the point where it isn't even economically viable to use such a precious resource. Thus, it's inevitable that fossil fuels will *need* to be replaced.

We obviously need to develop a replacement if fossil fuels are going to be phased out. We have a plethora of options already rising to prominence: Solar, wind, and hydroelectric are all ways we can generate energy based on the natural resources around us. Yet currently, these energy sources can't match the output of fossil fuels, which is a problem if we're going to need to replace fossil fuels. So, we need to work on developing these technologies, so they can soon be reliable enough to support everything we do. Or we consider other sources, like nuclear energy. Nuclear energy is both renewable (though some argue that it is nonrenewable because it requires uranium, a finite resource) and produces more energy than any type of fossil fuel. With technology becoming more energy-intensive, like AI, something as powerful as nuclear energy seems like the way of the future.

Yet, even if we could do all of that, it wouldn't remove the carbon dioxide that's already in our atmosphere, keeping our planet a greenhouse. All we've done is prevent the problem from getting worse. We haven't solved it. Therefore, we need to invest in carbon sequestration: processes that remove the carbon dioxide in our environments and the atmosphere.

One major process is by conserving and expanding the natural carbon sinks we have on land. The most obvious way we can do this is by stopping deforestation and promoting reforestation. We can't hope to remove the carbon dioxide from our atmosphere if we destroy the organisms and regions of land that are perfectly optimized for absorbing it. Therefore, we need to make efforts to conserve our forests. Once we are able to end deforestation, only then can we restore areas that once used to be forested. Carbon sequestration increases based on the age of a forest; the longer a forest is around, the more time it spends absorbing carbon. Trees are able to grow in size, and the microorganisms in the soil are able to break down the dead leaves and wood into more carbon, which is stored in the soil. Expanding our forests is also essential given that the efficiency of carbon absorption increases with the size of forests.

In the ocean, our carbon sinks take the form of seaweed and phytoplankton. Just like conserving and expanding our forests on land, growing our seaweed forests helps absorb carbon. It also keeps our oceans well-oxygenated, since seaweed emits oxygen that goes right into the ocean to prevent anoxia. However, the subject of phytoplankton is a tricky one. On the one hand, increasing phytoplankton populations will absorb carbon. But increasing them too much will create algal blooms, which do more harm than good. With global warming causing more erosion and providing more nutrients, it may seem like phytoplankton populations may spiral out of control and create widespread anoxia. Research has shown that being specific about what phytoplankton are fed and how much can help us avoid this problem. Erosion usually provides plankton with an abundance of nitrates, as this is what's in topsoil, and they're more concentrated in agricultural areas. This runoff is usually low in iron, which is needed to promote the production of chlorophyll, the compound that enables photosynthesis and carbon absorption. Therefore, by adding specific doses of iron to phytoplankton and monitoring their growth, we can promote populations of healthy sizes that can efficiently store carbon without leading to anoxia.

More ambitious efforts involve geologic processes. There is technology that can actually directly capture carbon dioxide from the atmosphere and compress it into a liquid. From there, it can be fed underground into empty oil and gas reservoirs where it can be stored in stable conditions for millions of years. It won't form back into fossil fuels, given that carbon dioxide is a byproduct of fossil fuel burning rather than a prerequisite for its formation. Instead, the carbon dioxide will slowly combine with the other chemicals in these geologic formations to form minerals like calcite, magnesite, and/or siderite.

Maybe the simplest and most beneficial way to remove carbon dioxide would be to incorporate it into our agriculture. There are several simple methods that work wonders in terms of storing carbon. One is composting, which takes dead plant material, fruit and vegetable waste, and occasionally livestock manure, and turns it into healthy, fertilized soil. It works great not just because it promotes healthy microorganisms in the soil that naturally absorb carbon dioxide, but also because it acts as a natural fertilizer that helps crops grow and increases their yield. Another is mulching, which involves spreading a layer of carbon-based material (like wood chips, straw, or even some types of shredded paper) over soil. Like composting, it serves dual purposes; the first is that topsoil is prevented from drying out, which keeps the microorganisms in the soil healthy and helps with water retention. The second is that the healthy soil and water retention is exactly what crops need to grow more food. Another is applying cover crops after we harvest. Rather than keep fields empty and lifeless that end up making soil less healthy, these fields can be planted with nitrogen-fixing plants in the offseason. Plants like buckwheat and white clover absorb nitrogen and store it in the soil, which helps microorganisms stay healthy and absorb carbon in return. It seems like a win-win situation; we get to remove carbon dioxide while also making the soil healthier for the plants we need to eat. If we adjust the way we grow crops to make it more sustainable, then we get around having to reduce farmland or change the types of food we grow—which are things we don't want anyway.

But the best part is that these are practices you can do. Composting, mulching, applying cover crops, and indeed simply making wise choices about the greenery you plant is entirely within your control. No voting necessary. No third-party services to contract out. In situations like this, it seems like grand policies or organizations are needed to take on such a massive issue. And they are needed—we need policies that focus on combating climate change, and organizations with vast resources that are capable of rebuilding carbon sinks. Yet, there's something to be said about the often-overlooked importance of simple decisions one person can make. These actions make a tiny difference compared to the gigatons of carbon dioxide being emitted, but it is a difference nonetheless. What happens when this scales? When thousands of people— or even millions of people—compost, mulch, use cover crops, and are selective in the plants they grow, we can make a serious difference.

Yes, we are facing some serious threats in our future. Unlike the dinosaurs, we can change our future. Earth always found a way to balance itself, even under the most horrifying circumstances. Yet it took millions of years for the climate to change, and dinosaurs were forced to adapt. We, on the other hand, are capable of leveraging the tools and processes that happen *naturally* to stabilize our planet in a much shorter time. In the process, we can make the world not just a better place, but one that is optimized for humanity.

Further Reading

This book stands on the shoulders of thousands of paleontologists and scientists who have worked over the past two hundred years to reveal all of this knowledge. This work is the culmination of hundreds of research papers and references. I'm truly thankful for the work of this entire field.

Instead of providing a long bibliography in this book, here is a website where you can download the complete reference list and the data that was studied for this book. Additionally, as a thank you for purchasing this book, you'll also receive a free one-month trial to Daily Dino Direct—an exclusive membership where you can dive deeper into paleontology and go from a dinosaur enthusiast to a dinosaur authority:

dailydinoguy.com/references

I would like to highlight a few pieces of literature for further reading; these studies in particular were a large influence on me when writing this book and would be a great starting place to dig deeper into the topics in this book.

Alvarez, Walter. *T. Rex and the Crater of Doom.* Princeton University Press, 2008.

Chiarenza, Alfio A., Juan L. Cantalapiedra, Lewis A. Jones, et al. "Early Jurassic Origin of Avian Endothermy and Thermophysiological Diversity in Dinosaurs." *Current Biology* 34, no. 11 (2024): 2517-2527. https://doi.org/10.1016/j.cub.2024.04.051.

Judd, Emily J., Jessica E. Tierney, Daniel J. Lunt, et al. "A 485-Million-Year History of Earth's Surface Temperature." *Science* 385, no. 6715 (2024). https://doi.org/10.1126/science.adk3705.

Scotese, Christopher R., Haijun Song, Benjamin Mills, et al. "Phanerozoic Paleotemperatures: The Earth's Changing Climate During the Last 540 Million Years." *Earth-Science Reviews* 215 (2021): 103503. https://doi.org/10.1016/j.earscirev.2021.103503.

Upchurch, Paul, and Alfio A. Chiarenza. "A Brief Review of Non-Avian Dinosaur Biogeography: State-of-the-Art and Prospectus." *Biology Letters* 20, no. 10 (2024): Article 20240429. https://doi.org/10.1098/rsbl.2024.0429.

Xu, Chi, Timothy A. Kohler, Timothy M. Lenton, Jens-Christian Svenning, and Marten Scheffer. "Future of the Human Climate Niche." *Proceedings of the National Academy of Sciences* 117, no. 21 (2020): 11350-11355. https://doi.org/10.1073/pnas.1910114117.

Want to learn even more about dinosaurs, paleontology, and science? Scan the QR code below to sign up for my newsletter and stay up to date on new dinosaurs being discovered, new dinosaur research, and more!

Acknowledgments

First and foremost, I would like to thank God for giving me the opportunity to write this book and glorify his creation in the process.

I would like to thank my wife, co-parent, and business partner, McKenzie. Her encouragement has been essential in keeping me going while writing something of this scale. She also does many of the unseen tasks that are involved in keeping a business running, which provided me with exceedingly valuable time to write this book.

I would like to thank Mango Publishing for giving me the opportunity to write this book. Synthesizing all of this research into one coherent story has been one of the deepest learning experiences of my life, and I hope you, the reader, learn as much as I did.

I would like to thank the many mentors who have shaped me into the scientist and communicator I am today. Thank you, Dr. David A. Burnham and the KU Paleontology Lab, for developing me during my bachelor's program. Thank you, Dr. Lindsay Zanno, Dr. Terry Gates, and the entire NCMNS lab, for developing me during my master's program. And thank you, Scott Dietz, for mentoring me and helping develop Daily Dino Guy.

I would like to thank all of my Daily Dino Direct members. Your support was essential in allowing me to write this book. I would also like to thank all of my fans on social media who follow and support me. Your engagement and passion for paleontology helped me get recognized for this opportunity.

Finally, I would like to thank the many researchers and professors in the paleontology field. Many people may ask why paleontology is important, and to that I say that each study is a crucial piece of the grand puzzle of our ancient past. Collectively, your work has ushered in a golden age of paleontology; we are able to reveal a more detailed vision of our past and begin to ask some truly amazing questions that seemed impossible to answer when I was a kid. I am more excited for the growth of this field as time goes on.

About the Author

Evan Jevnikar is a paleontologist and science communicator. Born and raised in Arizona, Evan moved to Kansas to receive his bachelor's degree in geology and then moved to North Carolina to receive his master's degree in biology. His master's thesis research was on the growth dynamics and life history of T. *rex*'s cousin *Tarbosaurus*. During his academic career, Evan excavated fossils in Kansas, Montana, New Mexico, and Utah. After graduating with his master's degree, he started Daily Dino Guy, whose mission is to bring the science of paleontology to life in a way that sparks understanding, develops trust in the scientific community, and invites everyone to explore Earth's ancient past.

mango
PUBLISHING

Mango Publishing, established in 2014, publishes an eclectic list of books by diverse authors—both new and established voices—on topics ranging from business, personal growth, women's empowerment, LGBTQ studies, health, and spirituality to history, popular culture, time management, decluttering, lifestyle, mental wellness, aging, and sustainable living. We were named 2019 and 2020's #1 fastest growing independent publisher by *Publishers Weekly*. Our success is driven by our main goal, which is to publish high-quality books that will entertain readers as well as make a positive difference in their lives.

Our readers are our most important resource; we value your input, suggestions, and ideas. We'd love to hear from you—after all, we are publishing books for you!

Please stay in touch with us and follow us at:
Facebook: Mango Publishing
Twitter: @MangoPublishing
Instagram: @MangoPublishing
LinkedIn: Mango Publishing
Pinterest: Mango Publishing
Newsletter: mangopublishinggroup.com/newsletter

Join us on Mango's journey to reinvent publishing, one book at a time.

* 9 7 8 1 6 8 4 8 1 8 6 9 3 *